# DIRECT TRANSISTOR-LEVEL LAYOUT FOR DIGITAL BLOCKS

# DIRECT TRANSISTOR-LEVEL LAYOUT FOR DIGITAL BLOCKS


PRAKASH GOPALAKRISHNAN
Neolinear, Inc.

ROB A. RUTENBAR
Carnegie Mellon University

**Kluwer Academic Publishers**
Boston/Dordrecht/London

Distributors for North, Central and South America:
Kluwer Academic Publishers
101 Philip Drive
Assinippi Park
Norwell, Massachusetts 02061 USA
Telephone (781) 871-6600
Fax (781) 871-6528
E-Mail <kluwer@wkap.com>

Distributors for all other countries:
Kluwer Academic Publishers Group
Post Office Box 322
3300 AH Dordrecht, THE NETHERLANDS
Telephone 31 78 6576 000
Fax 31 78 6576 474
E-Mail <orderdept@wkap.nl>

Electronic Services <http://www.wkap.nl>

**Library of Congress Cataloging-in-Publication**

Title: Direct Transistor-Level Layout for Digital Blocks
Author (s): Prakash Gopalakrishnan and Rob A. Rutenbar
ISBN: 1-4020-7665-7
ISBN: 1-4020-8063-8 (eBook)

# *Table of Contents*

# *Preface*

Cell-based design methodologies have dominated layout generation of digital circuits. Unfortunately, the growing demands for transparent process portability, increased performance, and low-level device sizing for timing/power are poorly handled in a fixed cell library. This motivated our search for an alternative layout technique, which has resulted in the direct transistor-level layout approach that we describe in this book. This new approach better accommodates demands for device-level flexibility for small blocks of custom digital logic. It captures essential shape-level optimizations, yet scales easily to netlists with thousands of devices, and incorporates timing optimization during layout.

This book would not have been possible without the support from several of our colleagues, friends and family. While we cannot mention all of them, we are particularly grateful to our spouses Surya Viswanathan, and Martha Baron; Prof. Larry Pileggi & Prof. Rick Carley of Carnegie Mellon University and Dr. Jeff Burns of IBM for their guidance and feedback throughout the research work; John Cohn & Dave Hathaway of IBM and Bill Halpin & Artour Levin of Intel for valuable discussions; Prof. Herman Schmit, Vikas Chandra, Aneesh Koorapaty, Emrah Acar, Ravishankar Arunachalam and Anoop Iyer for help with various experiments; colleagues and friends from Neolinear for their understanding and support. Finally, we thank the Semiconductor Research Corporation (SRC) for having provided the financial support for this work.

PRAKASH GOPALAKRISHNAN
NEOLINEAR, INC.


ROB A. RUTENBAR
CARNEGIE MELLON UNIVERSITY

# CHAPTER 1

# Introduction

## 1.1 Motivation: Standard Cells vs. Transistors

Standard-cell based design methodologies have dominated layout generation of digital VLSI circuits. These methods use a library of standard cells, which are pre-characterized layouts for circuits with about one to one hundred transistors. Standard cells have several notable virtues: they hide the increasingly unpleasant details of shape-level design rules; they arrange IO pins on individual gates in geometrically accessible locations; they assemble with relative ease into row-based blocks; they can be pre-characterized for timing and power, providing useful abstraction at higher levels like logic synthesis. As a result, research in layout automation has split into two directions: (a) device-level layout for individual cells; (b) cell-level placement and routing for large digital blocks. Figure 1 illustrates how cell-level tools arrange cells at block-level, while special techniques are used to generate device-level layouts for each of the cells from the standard-cell library.

**FIGURE 1.** Standard-cell based methodology for automatically generating layouts for digital blocks.

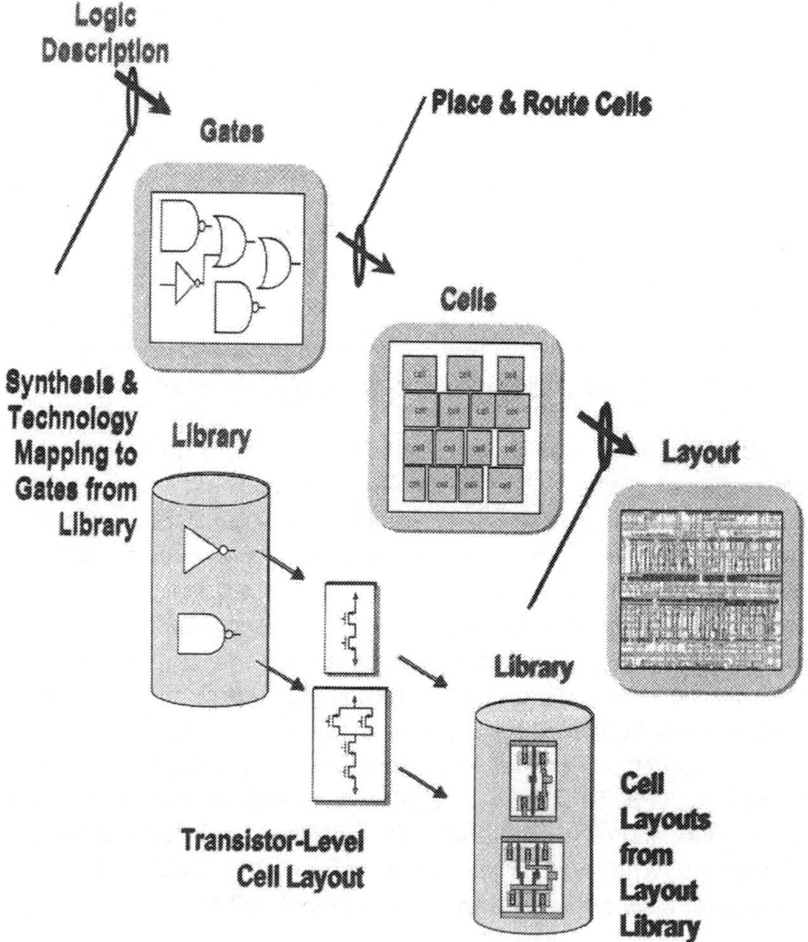

Unfortunately, the growing demands for transparent process portability, increased performance, and low-level device sizing to optimize block timing and power, are not easily handled in a fixed cell library. Libraries need to be large to achieve good logic synthesis results; today's best libraries comprise thousands of cell variants, enough to support multiple drive strengths and power/speed trade-offs. As a result, cell libraries carry an enormous inertia that resists porting, custom sizing, etc. This contributes to huge library maintenance costs.

**FIGURE 2.** Our direct transistor-level approach to generate layout.

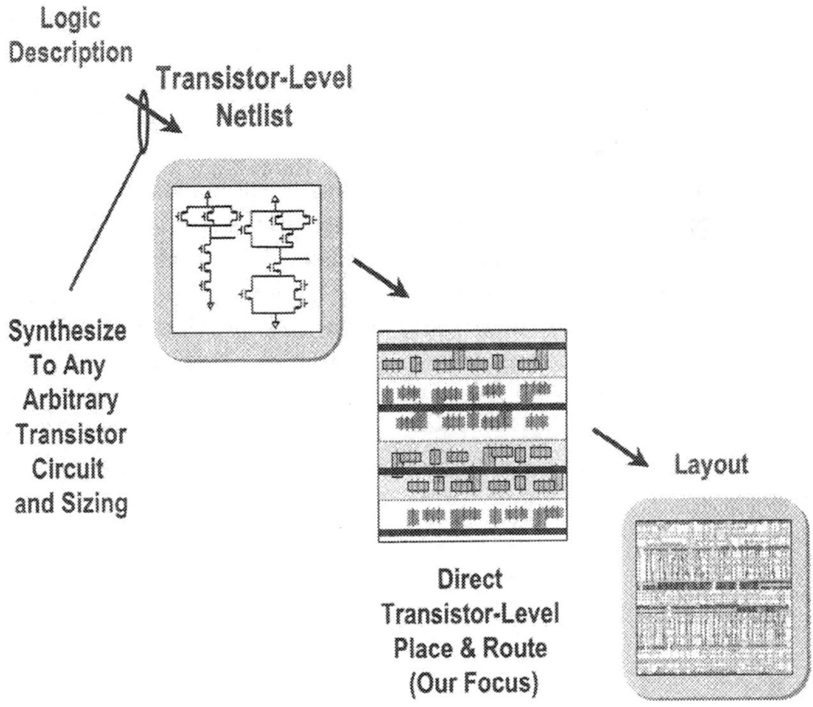

We propose a direct transistor-level layout approach (see Figure 2) for small blocks of custom digital logic as an alternative that can more easily accommodate the demands for device-level and shape-level flexibility. Unfortunately, the layout quality of transistor-level algorithms proposed to date leaves much to be desired. We argue that the essential flaw in these prior attempts is an over-reliance on the methods and assumptions of large-scale cell-based layout algorithms. Individual transistors may seem simple, but they do not pack as gates do for purposes of optimal layout. Careful inspection of the internal layout of any well-designed library cell will reveal a wealth of shape-level optimizations that each shave a bit of area or delay off the overall layout. These savings may seem negligible, but they are *amplified* enormously in any layout with thousands of cells. Algorithms that capture some of these geometric tricks usually rely on optimization frameworks that cannot scale beyond a few tens of devices. Algorithms that ignore these shape-level issues and pretend devices can be laid out as if they were gates suffer the consequences when thousands of devices are poorly packed.

The lack of alternatives to timing characterization has severely hindered adoption of previous transistor-level approaches in an industrial flow. Given the importance of interconnect delays during deep submicron (DSM) technologies, the effect of layout on timing closure cannot be ignored. Pre-characterized standard cells provide useful models for timing analysis during layout optimization. However, the recent emergence of efficient, accurate transistor-level timing estimators ([Dartu 98], [McDonald 01]) mitigates this problem. Some of these estimators are fast enough to be useful within a layout optimization framework.

## 1.2 Our Goals

In this book, we develop a novel set of algorithms for direct transistor-level layout that:

(a)capture the essential shape-level optimizations, yet

(b)scale easily to netlists with thousands of devices, and

(c)incorporate timing optimization during layout.

The key idea is early identification of essential diffusion-merged MOS device groups, and their preservation in an *uncommitted* geometric form until the very end of detailed placement. Roughly speaking, we extract essential groups *early* from the transistor-level netlist, place them *globally*, optimize them *locally*, and then finally *commit* each to a specific shape-level form while concurrently optimizing for both density and routability. This division of the netlist into essential clusters further makes it possible to analyze the timing-behavior of the circuit, which we use to guide placement optimization by accounting for estimated timing delays due to the interconnect. We use a commercial detailed router to complete our flow. Results to date are encouraging: we can place/route 2000 devices in about 30 minutes, and we can consistently save 15-30% on area in comparison with a standard commercial cell-based flow. Our timing optimization techniques achieve about 10-20% reduction in overall circuit delays.

We now review existing transistor-level approaches and also build the geometric foundation for our layout abstraction. We then present an overview of our methodology, highlighting its key differences from existing methods.

## 1.3 Previous Transistor-Level Approaches

Prior work on transistor-level digital layout focuses on two targets: layout for individual library cells, and layout for larger blocks of arbitrary logic. We briefly look at each of these.

In the context of cell layout generation, Maziasz [Maziasz 92] provides a good review of the graph theoretical formulations and algorithms for one-dimensional cell layout synthesis. Some approaches, like Malavasi [Malavasi 95], LiB [Hsieh 91] and PicassoII [Lefebvre 92], partition the circuit into locally optimal clusters of transistors and then place these clusters. However, it has been noted by Sadakane [Sadakane 95] and HCLIP [Gupta 99] that techniques which generate intra- and inter-cluster layouts in two separate stages can yield layouts that are far from optimal. The concept of geometric options for clusters during final placement (Sadakane [Sadakane 95]) is one we will revisit—though we present a novel scalable formulation. Although HCLIP does explore inter-cluster diffusion merging using a hierarchical ILP formulation, we believe that this must be exploited dynamically, in the context of overall physical placement and routability. Further, the maximum number of transistors that can be handled by any of these methods is about 100 to 200. We aim at circuits that are an order of magnitude larger.

Basaran [Basaran 97] begins by transferring a MOS circuit into a *diffusion graph*. Formally, this graph has a node for each net and an edge for each MOS device channel. Any trail through the graph identifies a series of devices, all of whose drain/source diffusions may be shared if the devices are placed in a row, in the same order the path visits the edge associated with each device (see Figure 3). It was shown that iterative improvement methods involving trail reordering can then be used to search the solution space of min-width layouts for those that minimize routing. The work by Basaran is one we will re-visit in our discussions on shape-level optimizations. Riepe [Riepe 99] uses Basa-

**FIGURE 3.** Trails in Diffusion Graph and Layout

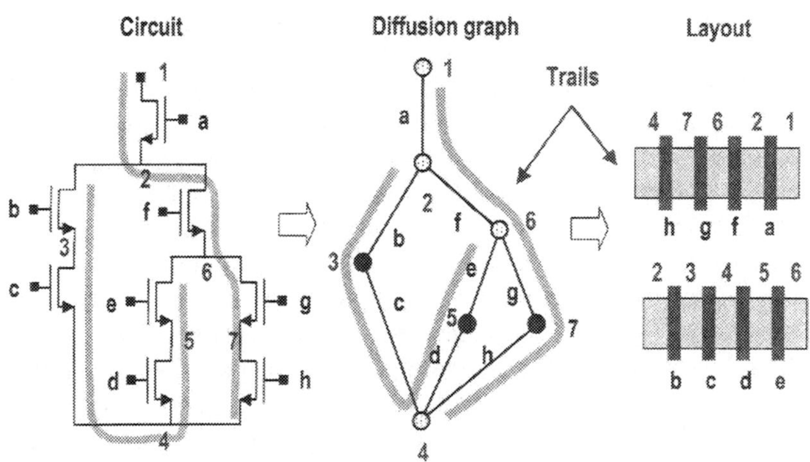


ran's techniques to dynamically alter trails within a 2-D placement using a sequence pair formulation. More recent work, AKORD [Serdar 99], uses pre-calculated optimal chaining orders to accept/reject random moves within an annealing framework, in the style of KOAN [Cohn 94]. These chains of transistors, called *trails*, are a natural abstraction for layout optimization. However, as before, these algorithms do not scale well for larger circuits.

It has been noted by Lefebvre and Chan [Lefebvre 89] that for 2-D layouts, a solution with minimum wiring complexity is not necessarily of minimum width. This leads us to believe that constraining transistors too early into dense, min-width, locally optimal, clusters can hamper placement flexibility.

In the realm of block-placement at transistor level, Tani, *et al.* [Tani 91] partition the circuit into sub-circuits using a temporary cell library which they create. Each sub-circuit is a collection of cells in a row. They then optimize each row for one-dimensional placement. They have generated layouts with areas smaller than standard cell-based layouts—but only in an older channel-based routing style. Their saving came from allowing routing over the cell, which is the norm today.

Three recent industrial cell and block synthesis tools address related practical problems. Cellerity [Guruswamy 97] employs annealing-based optimization for diffusion abutment, wirelength, channel density and gate alignments during leaf cell synthesis. C5M [Burns 98] develops hierarchical row-based macros for static CMOS logic. Schematic partitioning is combined with device-size-tuning and on-the-fly leaf-cell synthesis, accommodating multiple objectives and top-down constraints. However, our method aims to eliminate explicit cell generation. Nevertheless, we note that pin positioning and cell image flexibility are crucial to routability. LAS [Chow 92] produces flat symbolic layouts at block-level, which are then compacted. A major drawback is the preliminary partitioning into locally optimal clusters which are not thereafter dynamically altered in the context of the overall placement. We have observed empirically in our own tools that dense diffusion merging without careful routability consideration will require excessive white space to accommodate routing. This results in sub-optimal overall layouts.

In the next section, we introduce our new strategy that overcomes the quality and scalability problems of these previous approaches. Further, our method incorporates timing optimization that is lacking in these prior transistor-level approaches.

## 1.4 Overall Strategy

We begin with a fundamental question: *in a transistor-level layout tool, what exactly are we placing and routing?* Clearly, standard-cells offer a rather coarse abstraction as fundamental units of placement. Individual transistors, on the other hand, tend to be too small, especially if we want both shape-level optimization (which is very local) and accurate routability prediction (which requires a more global view). As mentioned in the previous section, *trails* (see Figure 3), which are chains of diffusion-merged transistors, are natural device groupings. We later show that by abstracting the netlist as groups of trails, we can effectively support geometric optimization from shape-level up to block-level.

**FIGURE 4.** Our Geometric Model

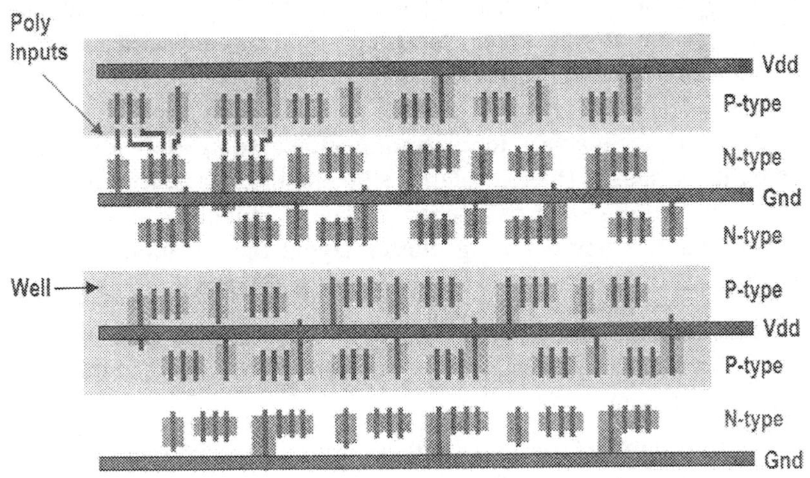

For the static CMOS style, we restrict ourselves to a row-based geometric model that resembles a standard cell-based footprint. As illustrated in Figure 4, we have two rows of N-type trails alternating with two rows of P-type trails. There are two reasons for this choice: (a) transistors from alternating rows of the same diffusion type can share a common power rail and also the same well region; (b) transistors from neighboring rows of opposite diffusion type can have aligned polysilicon gates.

Transistor clustering has been a key step in many prior layout algorithms. However, the aggressive clustering needed to handle flat placement of thousands of devices invariably leads to locally-optimal, but globally sub-optimal groups that limit the shape-level optimizations we want to explore. We resolve this by initially forming only *essential* clusters of devices, which are mandatory for the given circuit style. For example, for static CMOS circuits, these are transistors that are strongly connected and constrained to be together due to polysilicon gate-alignment. A cluster is represented as a group of connected transistor trails. In our approach, "global" optimization focuses on these clusters as the atomic units; "detailed" optimization focuses on the trails inside a specific cluster.

Previous approaches to clustering either completely disallowed inter-cluster merges (like standard-cell methods), considered inter- and intra- cluster merges in two separate stages, or ignored global placement/routability issues during cluster layout optimizations. The key components of our trail-level approach are as follows.

- **Minimal Essential Clustering**: We initially form essential clusters to meet circuit style-imposed layout restrictions. We keep this minimal for maximum flexibility.

- **Circuit Structure Recognition**: We recognize structure in transistor netlists and store distinct circuit structures in a library for pattern matching. These circuit patterns are our basis for clustering. This creates a level of hierarchical divide-and-conquer that lets us effi-

**FIGURE 5.** Flow Overview

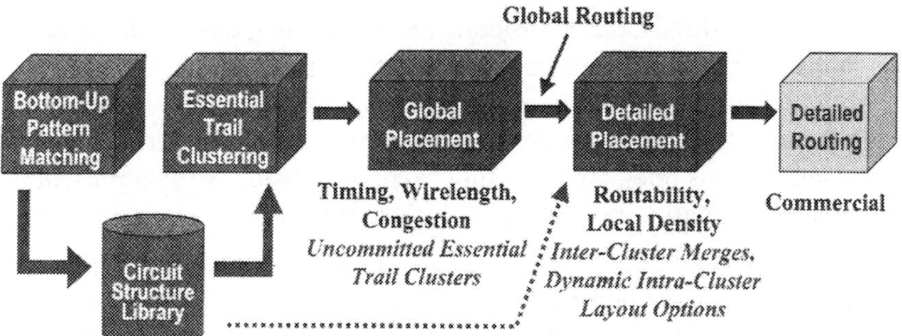


ciently compute all feasible trail-level layouts for any cluster (pattern). This is efficient because we only need to compute essential trail formations for a few distinct circuit structures.

- **Timing-Driven Global Placement**: We find optimal locations for our essential clusters while minimizing overall wirelength, congestion and timing delays. The clusters, however, are not committed to any fixed layout option (any arrangement of transistor trails) during this phase.

- **Detailed Placement**: We next explore dense inter-cluster diffusion merging considering all feasible intra-cluster layout options, in the context of global placement and routability. This is where each cluster is finally bound to a physical shape-level trail layout. Global routing is updated dynamically during this optimization.

To complete the flow, we have integrated a commercial detailed router.

## 1.5 Book Outline

The remainder of this book describes our algorithms in detail and is organized as follows.

- Chapter 2, titled Circuit Structure and Clustering, describes our techniques for circuit structure recognition and essential clustering, which form the basis of our shape-level optimizations and overall placement. We describe here the benchmarks used for various experiments in this book and how they were generated. We also present results comparing benchmarks synthesized using different libraries.

- Chapter 3 describes our global placement algorithms. For simplicity, we only describe non-timing driven algorithms here. Timing optimization is discussed in detail separately. Results demonstrating influence on congestion are presented.

- Chapter 4 describes our detailed placement techniques and summarizes our routing integration. We present fully routed layout results comparing our flow with a standard cell-based flow.

- Chapter 5 describes our timing optimization during global placement. We present results comparing our approaches with and without timing-optimization.

- Chapter 6 offers concluding remarks and future work.

We have implemented our ideas in a tool called *TrailBlazer*. The results presented in each of the chapters used TrailBlazer to validate the relevant algorithmic aspects of our flow.

# CHAPTER 2

# Circuit Structure And Clustering

## 2.1 Introduction

Our transistor-level approach abstracts the netlist as clusters that are groups of strongly connected trails or chains of transistors. Compared to individual transistors, clusters make it possible to tackle netlists with thousands of transistors, while enabling geometric optimization from shape-level up to block-level. For maximum placement flexibility and to prevent clustering from generating locally optimal but globally sub-optimal groups, we only form essential clusters of devices, which are mandatory for a given circuit style. These restrictions include constraints that require groups of transistors forming logic stages, to be laid out in a particular fashion, adjacent to each other. As we shall see in a later chapter, such a clustering of groups of transistors is also necessary to enable timing analysis of the circuit.

To motivate our choice of groups of trails as clusters, we begin this chapter with a discussion of trails and their usefulness in transistor-level layout. We then describe how we exploit inherent local structure in netlists to form essential clusters and efficiently explore shape-level optimization.

## 2.2 Trails

Trails are chains of transistors that are merged at their source/drain terminals in the layout. In order to appreciate their utility in shape-level layout optimization, we review here the work by Basaran [Basaran 97].

**FIGURE 6.** Example Eulerian Trail: (a) the circuit partition, (b) the diffusion graph, super-edges are shown with dashed lines, (c) an Eulerian trail and the corresponding trail cover

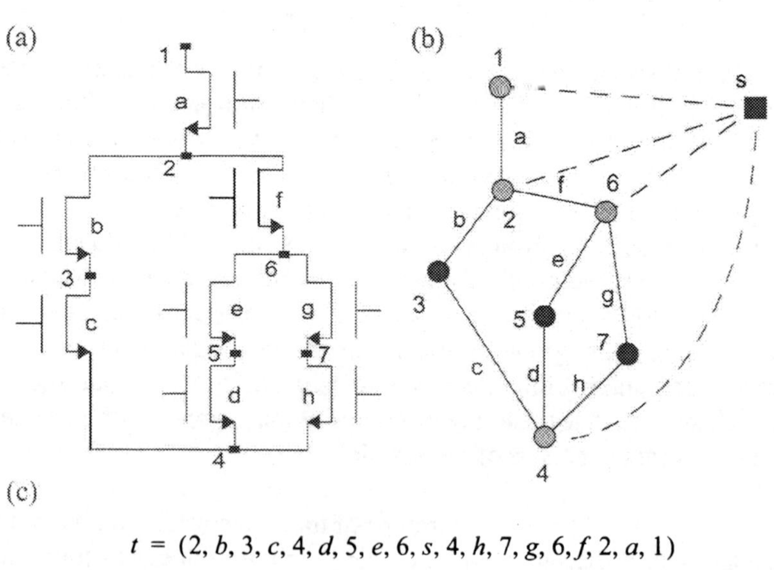

$$t = (2, b, 3, c, 4, d, 5, e, 6, s, 4, h, 7, g, 6, f, 2, a, 1)$$

$$T = \{(2, 3, 4, 5, 6), (4, 7, 6, 2, 1)\}$$

Basaran begins by transferring a MOS circuit into a *diffusion graph*. Formally, this graph has a node for each net and an edge for each MOS device channel. Hence, the graph only has edges between nodes that represent transistor drains and sources. The essential utility of the diffusion graph is that any long path through the graph identifies a series of devices, all of whose drain/source diffusions may be shared if the devices are placed in a row, in the same order the path visits the edge associated with each device. Figure 6(b) shows the diffusion graph for the circuit in Figure 6(a).

Ideally, we would like to be able to find a path which includes *all* devices to be placed in a row and merged. Such a path through the graph is called an *Eulerian path* or *Eulerian trail*. However, such a path does not always exist. In particular, it cannot exist if any node in the graph has an odd degree. Basaran showed that by adding a single distinguished node to the graph, called a *super-vertex* and edges from each odd-degree node to this super-vertex (*super-edges*), the modified diffusion graph has an Eulerian trail. He also showed that the super-edges correspond to *gaps* in the placement of the devices in a row. Also, *any* Eulerian trail in this augmented graph maximally merges the MOS devices when placed in a row, and minimizes the number of gaps where adjacent MOS devices have drains/sources which cannot be shared. Figure 6(c) shows an Eulerian trail corresponding to the diffusion graph in Figure 6(b). The corresponding layout of minimum width is shown in Figure 7.

Basaran uses this idea to suggest a novel optimization strategy for the simple single-row MOS layout problem. Instead of representing the geometry of the devices explicitly, he manipulated Eulerian trails in the diffusion graph. The idea was to do iterative improvement to reshape the Eulerian trail, thus creating a different one-dimensional placement. The idea is also to explore *only* in the space of minimum-width single-row layouts, using as objective function, the number of wiring tracks needed to route the layout.

**FIGURE 7.** Basaran's techniques for getting different diffusion merges: Stacked layout for Figure 6 (c)

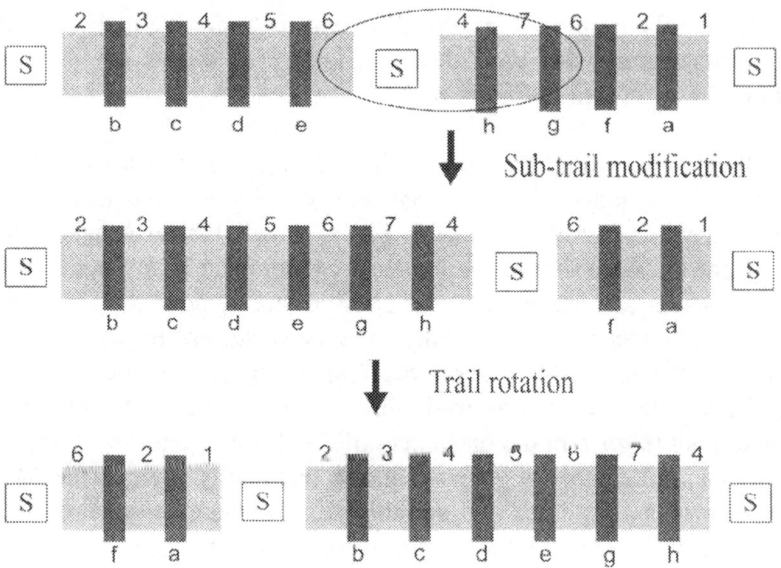


The solution space of all layouts of minimum width is traversed using simulated annealing with two annealing moves. A **sub-trail modification** move picks, at random, a sub-trail within the layout and generates a new sub-trail randomly with the same terminals at its ends. This is achieved by evaluating a new Eulerian trail in a sub-graph formed by the elements of the sub-trail. A **trail rotation** move rotates the entire trail so that the relative ordering of the sub-trails in the Eulerian trail change. These two moves have been proven to be complete [Basaran 97]. Figure 7 illustrates these for the example from Figure 6.

Basaran's techniques illustrate an important point that trails form natural groupings of transistors. Further, they also demonstrate that special techniques are required to explore trail formations for local groups of transistors. Unfortunately, Basaran's techniques do not scale beyond netlists with few tens of transistors. In the next section, we explain how circuit-style imposed constraints further restrict trail formation possibilities for local groups of transistors. We then describe how we exploit circuit structure to tackle problems with a large number of transistors, while incorporating trail-level optimizations.

## *2.3 Essential Clusters and Circuit Structure*

Most transistor netlists and their corresponding layouts have a significant amount of important local structure. When style-specific circuit constraints are imposed, there are not too many ways in which the transistors can be combined to produce routable layouts. For example, an important requirement for series-parallel static-CMOS circuit layouts is the alignment of polysilicon gates belonging to complementary transistors. Another performance-induced constraint is the level of strapping required on the source/drain diffusion nodes. Guan et. al. [Guan 95] describe techniques for creating layouts when partial/full diffusion strapping is required. It turns out that even for the most commonly occurring circuit structures, there may not be several different ways of combining the transistors to form trails while paying attention to these details. For the circuit shown in Figure 8, there is only *one* such way of combining the transistors so that the resulting placement is routable. For some other simple circuits like that in Figure 9 there are more trail formation options to choose from.

These complementary pairs of trails together form the *essential trail clusters* that *need* to be together because of circuit style imposed restrictions. We refer to the corresponding layout options for these clusters as *cluster layout options*. Notice however that these options are still

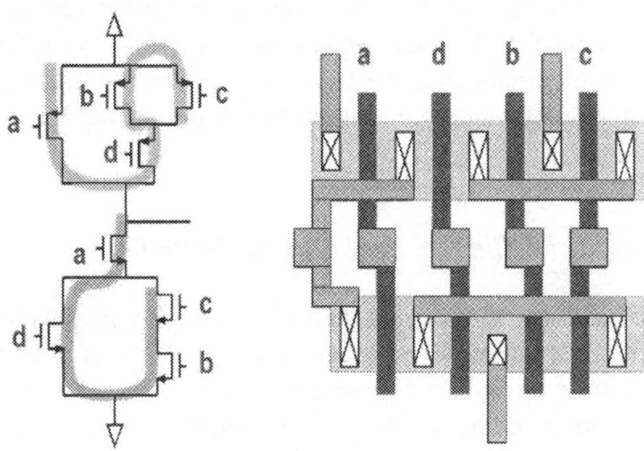

**FIGURE 8.** Circuit with only one configuration for trails

modest in number (two for the simple NAND example in Figure 9). Consequently, we argue that the cluster layout options should not be discovered on-the-fly, but should instead be dynamically re-arranged during placement.

Further, in large transistor-level netlists, there are often repetitive circuit structures. This is especially true of control logic and data-path circuits that are commonly synthesized by technology mapping to a (sized transistor-level) logic library. Our key idea is to identify these circuit structures, and group them, bottom-up via pattern matching. We describe this in the next few sections.

**FIGURE 9.** Simple NAND circuit with two possible configurations for trails.

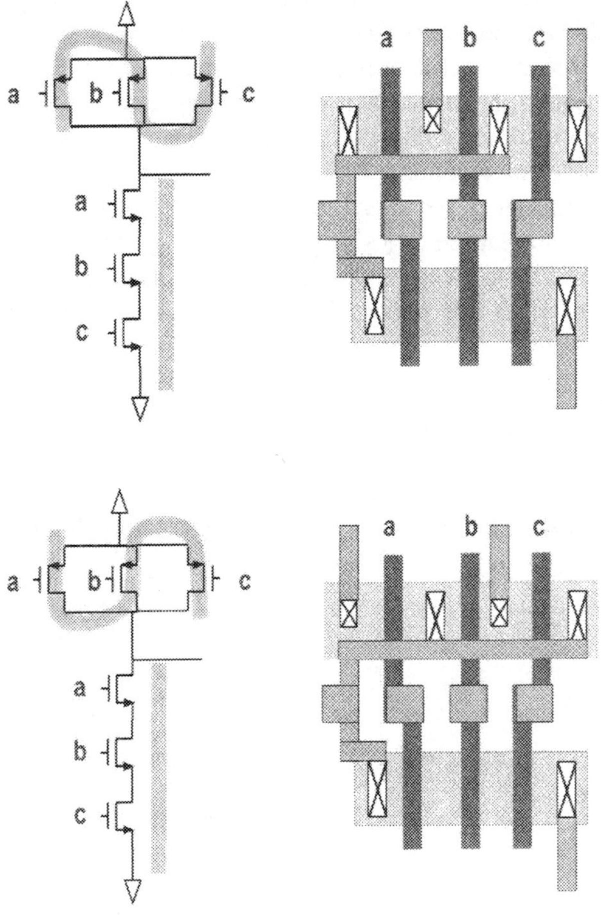

## 2.4 Pattern Matching

We use graph isomorphism techniques for pattern matching. There have been similar approaches to pattern matching in netlist partitioning [Ohlrich 93]. We make use of the Graph-Matcher tool library [Messmer 95]. Graph-Matcher maintains model graphs in its database (*Graph-MatcherDatabase*) that it uses for pattern-matching with input graphs via graph isomorphism. We start by generating a *circuit graph*, that has a one-to-one mapping with a given circuit structure. This is achieved by having a node corresponding to each source, gate and drain terminal of individual transistors. All source/drain nodes connected to the same net can further be collapsed into a single node. For gates connected to each other, we introduce a new node for the common net and connect each such gate node to this node. We also label the power nodes (VDD/

**FIGURE 10.** Example Circuit Structure Graph used for pattern-matching via graph isomorphism

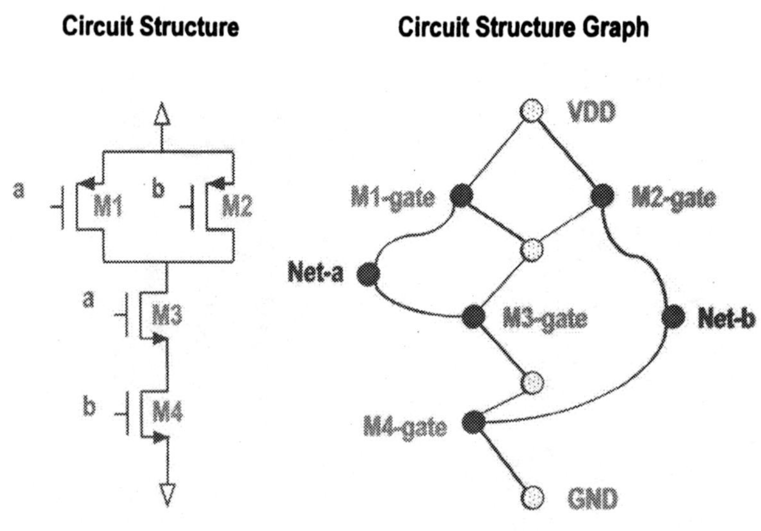

GND), diffusion nodes and gate nodes differently. Such a graph structure enables us to correctly match circuit structures while paying attention to the connectivity. An example circuit graph for a simple NAND circuit is shown in Figure 10.

Figure 11 shows our algorithm for pattern matching. As shown here, if there are any special circuit structures that the user cares about,

**FIGURE 11.** Algorithm for circuit structure recognition via pattern-matching.

---

| | |
|---|---:|
| **Initialize** *GraphMatcherDatabase* with no models | |
| **foreach** special circuit structure, $\underline{c}$ | 1 |
|     g = generate circuit graph for $\underline{c}$ | 2 |
|     Insert g as new model in GraphMatcherDatabase | 3 |
| end | 4 |
| **let** *Clusters* be essential circuit clusters from netlist | 5 |
| **foreach** circuit in *Clusters*, $\underline{c}$ | 6 |
|     g = generate circuit graph for $\underline{c}$ | 7 |
|     m = No. of matches of *g* with *GraphMatcherDatabase* models | 8 |
|     if *m > 0* | 9 |
|         Record $\underline{c}$ as instance of matched model circuit structure | 10 |
|     else | 11 |
|         Insert *g* as new model in *GraphMatcherDatabase* | 12 |
|         Record $\underline{c}$ as new circuit structure for the new model | 13 |
|     end | 14 |
| end | 15 |

these can be used to initialize the *GraphMatcherDatabase*. The netlist is partitioned into essential circuit clusters by identifying tightly coupled transistors. Any partitioning scheme can be used at this step. In our case, we simply identify the transistors that are tightly connected to each other via source/drain connections. Such groups are also referred to as dc-coupled groups or *DCCs*. The utility of DCCs will become clear when we discuss timing analysis in a later chapter. Note that this partitioning defines our minimal essential clustering. Each circuit cluster is then matched with the models in the *GraphMatcherDatabase* as shown. Note also that we build this database of models as we find new structures in the netlist. This lets us discover new structures in the netlist as well as identify structures that are repetitions of each other.

## 2.5 Circuit Structure Library

Having identified distinct circuit structures, we determine optimal trail formation options for these circuit structures and store them in a *Circuit Structure Library*. As shown in Figure 12, this provides a framework that allows us to use any of the known diffusion graph algorithms, like Basaran's, or other cell layout synthesis techniques to determine optimal trail options. This results in significant computational savings because we only need to determine optimal options for small, distinct circuit structures in the netlist.

Sadakane [Sadakane 95] uses clusters and template-options for clusters, but their method is extremely time-consuming and handles only up to 80 transistors. This could be attributed to their annealing-based intra-cluster option selection. In a later chapter, we explain our novel ideas for using the circuit structure library and trail options to explore inter-cluster diffusion merges between all feasible intra-cluster layout options, exhaustively and deterministically, in the context of global placement and global/detailed routability.

**FIGURE 12.** Circuit Structure Library

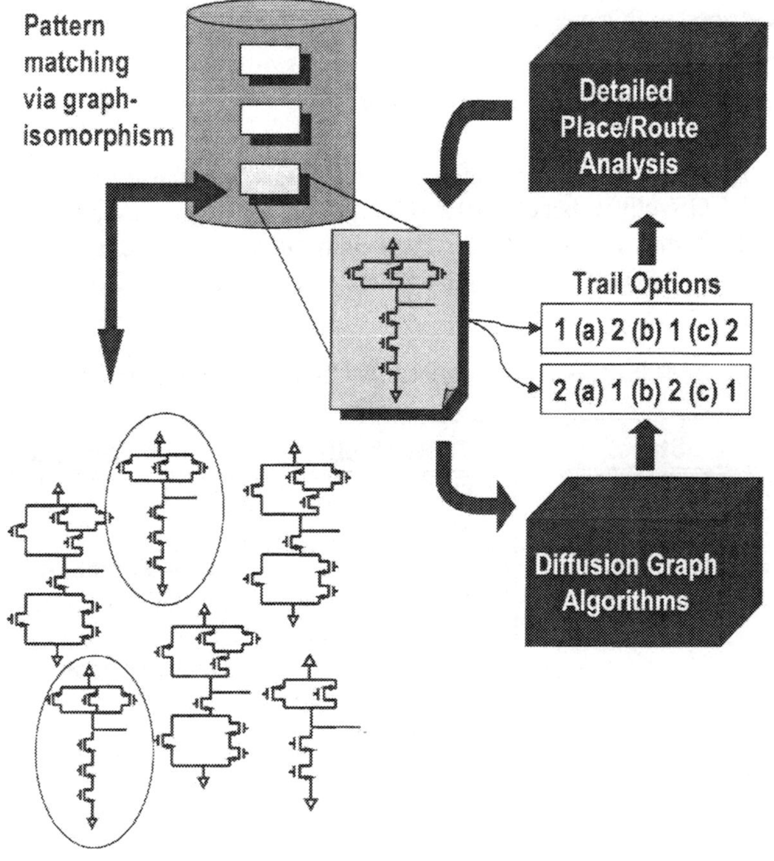

## 2.6 Benchmarks, Clustering Experiments and Results

In this section, we first discuss the benchmarks used for various experiments in this thesis and describe our process for generating them. We then present results comparing a set of benchmarks synthesized using different libraries, in the context of the clustering techniques described in this chapter.

### 2.6.1 Benchmarks for Experiments

For the various experiments in this thesis, we chose several standard LGSynth91 benchmarks [CBL]. The are logic descriptions (in BLIF formats). We compile these into functional schematics using logic synthesis, which we then flatten to a transistor-level netlist. For

**TABLE 1.1** Benchmark generation for HP0.35 μm technology

| Steps | Tools / Method | File Formats |
|---|---|---|
| Logic description | LGSynth91 benchmarks [CBL] | Input: BLIF |
| Logic synthesis | Using SIS [Sentovich 92] for logic synthesis. Simple target library containing INVERT/AND/NAND/OR/NOR gates up to 4 inputs. | Output: Functional schematic in NETBLIF format, using logic gates from target library |
| Transistor-level flattening | Using scripts and transistor-level implementations of library elements from target library. | Output: Transistor-level netlist in TrailBlazer's TRAN format |

specific algorithmic comparison, we have used technology parameters from either the 0.35um HP process or the 0.18um STMicroelectronics process. In general, the choice was based on the availability of certain parameters and infrastructure. For example, owing to the availability of the HCMOS8D CORELIB (full standard cell library) from STMicroelectronics, we used it for library comparison and clustering experiments presented in this chapter. Similarly, based on available CCT routing integration, the HP 0.35um technology parameters were used for layout comparison experiments. Timing-driven experiments are

**TABLE 1.2** Benchmark generation for STMicroelectronics 0.18um technology

| Steps | Tools / Method | File Formats |
| --- | --- | --- |
| Logic description | LGSynth91 bench-marks [CBL] | Input: BLIF |
| Intermediate AND-OR logic | Using SIS [Sentovich 92] & scripts | Output: Verilog AND-OR logic |
| Logic synthesis | Using Design Compiler [Synopsis] for logic synthesis. Target libraries created using various subsets of HCMOS8D CORELIB from [STM] and compiled using Synopsis Library Compiler | Output: Verilog functional schematic using logic gates from target library |
| Transistor-level flattening | Using scripts and transistor-level implementations of library elements from HCMOS8D CORELIB (SPI format) | Output: Transistor-level netlist in TrailBlazer's annotated SPICE format |

presented using STMicroelectronics 0.18um parameters, owing to available wiring delay estimates and simulation tool parameters. Details of how the transistor-level benchmarks were generated from the original logic descriptions, for the HP & STMicroelectronics technologies, are provided in Table 1.1 & Table 1.2 respectively.

We now compare benchmarks using different combinations of synthesis target libraries for the STMicroelectronics 0.18 technology and present results from clustering.

### 2.6.2 Synthesis Target Library Comparison

We compare the results of logic synthesis for a set of five benchmarks, synthesized using different target libraries. We generated five target libraries of varying sizes, using the full HCMO8D CORELIB standard cell library and some arbitrarily chosen subsets, as shown in Table 1.3 .

Figure 13 shows the sizes of the various libraries in terms of the number of cells in the library and also its effect on the total number of gates in each of the five benchmarks. Note that as the library size increases, the number of gates in the netlists decreases, since synthesis has the option of choosing from complex cells. This is one of the advantages of having flexibility during synthesis, in the form of a larger library.

For the benchmarks synthesized above, Figure 14 compares the number of unique library cells that actually get used in the various netlists, with the increasing library sizes. Notice here that the number of unique cells that got used during synthesis is almost an order of magnitude smaller than the number of unique cells available in the library, as the library size increases.

TABLE 1.3 Elements in various libraries that are subsets of
HCMOS8D CORELIB

| Library | Elements |
|---|---|
| lib-0 | IVLL, ND2LL, ND3LL, ND4LL, NR2LL, NR3LL, NR4LL |
| lib-1 | **lib-0 elems** + AN2LL, AN3LL, AN4LL, OR2LL, OR3LL, OR4LL, AO10LL, AO12LL, AO13LL, AO14LL, AO15LL, AO16NLL, ... <br> *(See Appendix for full list)* |
| lib-2 | **lib-0 elems** + AN2LL, AN2LLP, AN2LLX3, AN2LLX4, AN3LL, AN3LLP, AN3LLX3, AN3LLX4, AN3LLX8, AN4LL, AN4LLP, AN4LLX3, AN4LLX4, IVLLP, IVLLX05, IVLLX16, IVLLX3, IVLLX32, IVLLX4, IVLLX5, IVLLX8, ND2LLP, ND2LLX05, ND2LLX3, ND2LLX4, ND3ALL, ND3ALLP, ND3ALLX3, ND3ALLX4, ND3LLP, ND3LLX05, ND3LLX3, ND3LLX4, ... <br> *(See Appendix for full list)* |
| lib-3 | **lib-2 elems** + AO10LL, AO10LLX05, AO10NLL, AO10NLLP, AO11LL, AO11LLP, AO11LLX05, AO11NLL, AO11NLLP, AO12LL, AO12NLL, AO12NLLP, AO13LL, AO13LLX05, AO13NLL, AO13NLLP, AO14LL, AO14NLL, AO14NLLP, AO15LL, AO15LLX05, AO15NLL, AO15NLLP, AO16LL, AO16NLL, AO16NLLP, AO17LL, AO17LLX05, AO17NLL, AO17NLLP, AO18LL, AO18LLX05, AO18NLL, ... <br> *(See Appendix for full list)* |
| lib-4 | Entire library |

**FIGURE 13.**  Number of gates in benchmarks for various library sizes

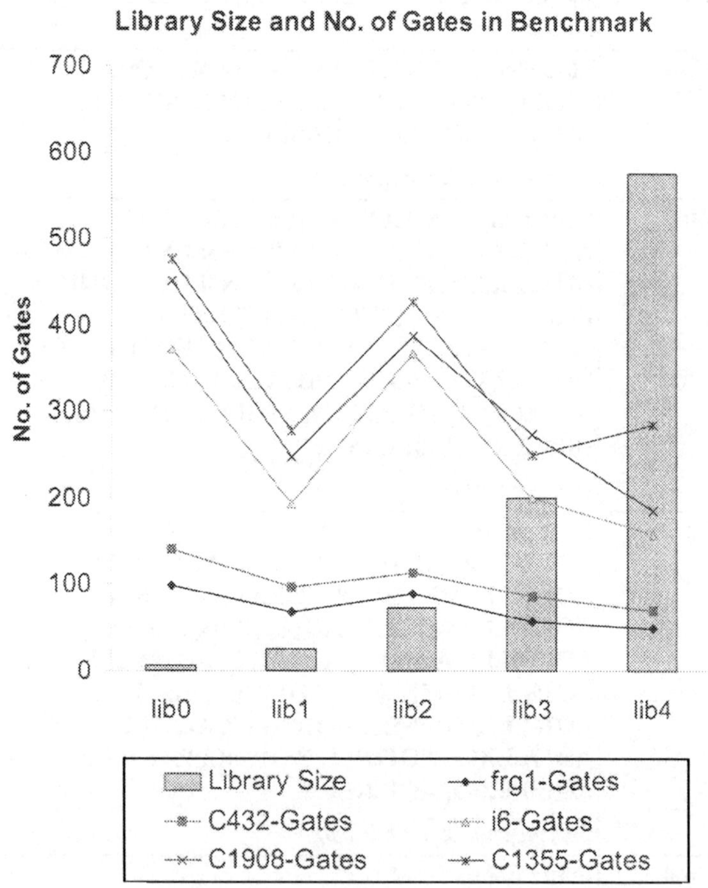

**FIGURE 14.** Number of unique cells used in benchmarks for various library sizes

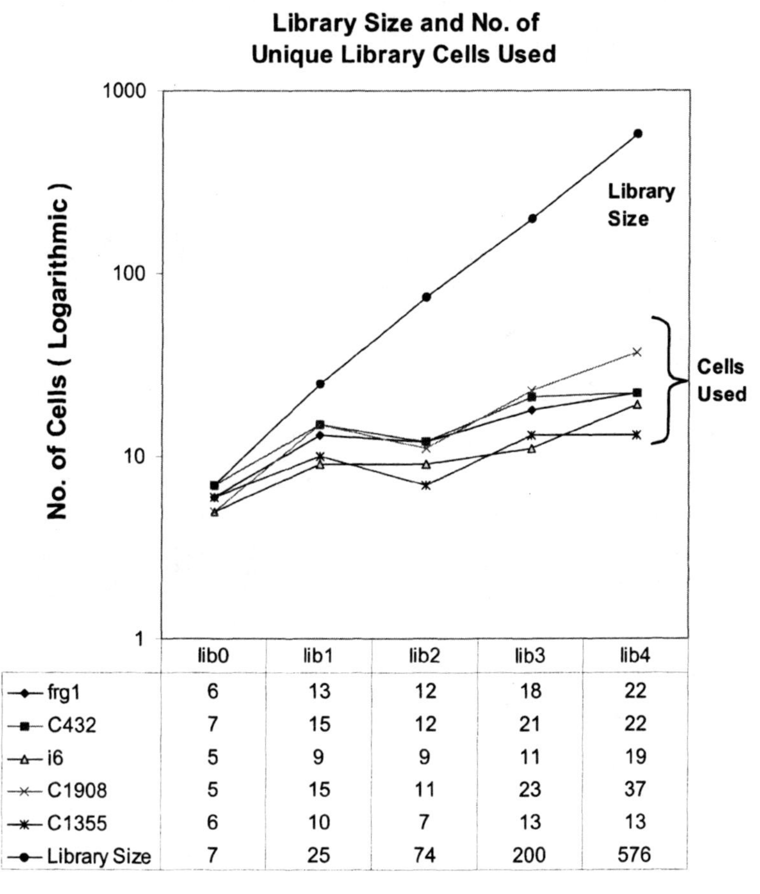



| | lib0 | lib1 | lib2 | lib3 | lib4 |
|---|---|---|---|---|---|
| frg1 | 6 | 13 | 12 | 18 | 22 |
| C432 | 7 | 15 | 12 | 21 | 22 |
| i6 | 5 | 9 | 9 | 11 | 19 |
| C1908 | 5 | 15 | 11 | 23 | 37 |
| C1355 | 6 | 10 | 7 | 13 | 13 |
| Library Size | 7 | 25 | 74 | 200 | 576 |

**FIGURE 15.** Comparing unique cells used with unique circuit structures in the benchmarks, across various libraries.

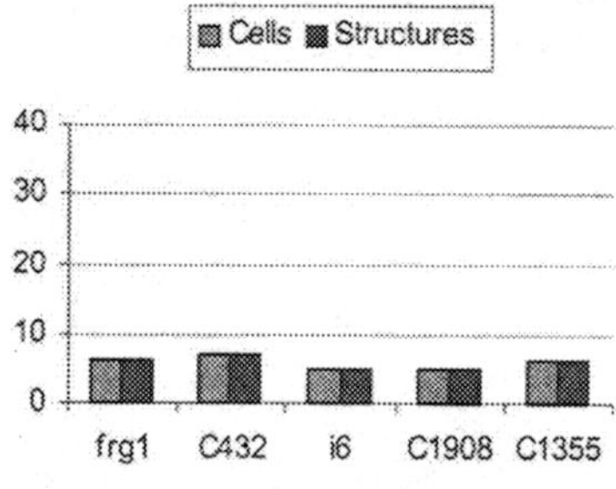


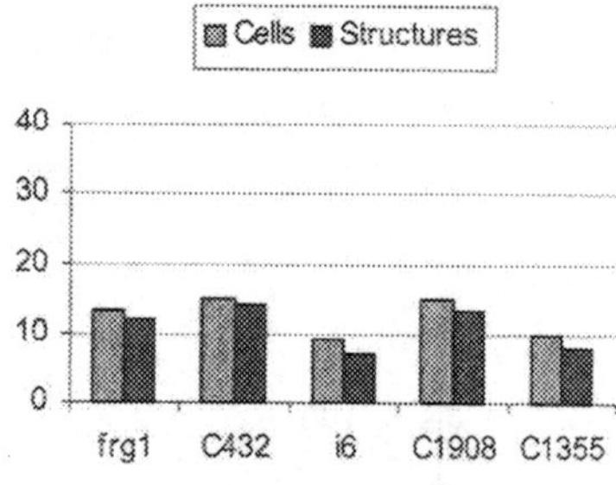


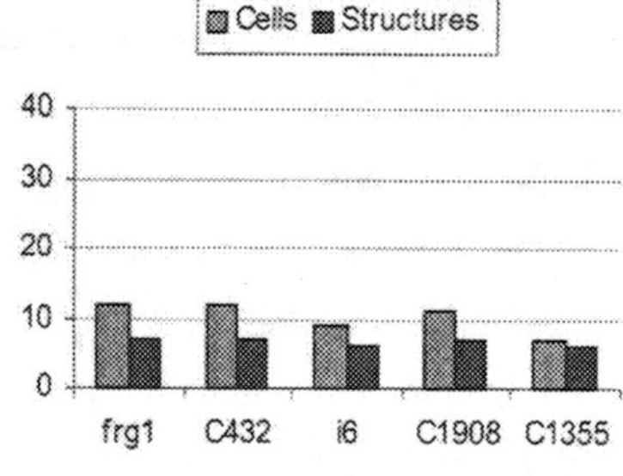


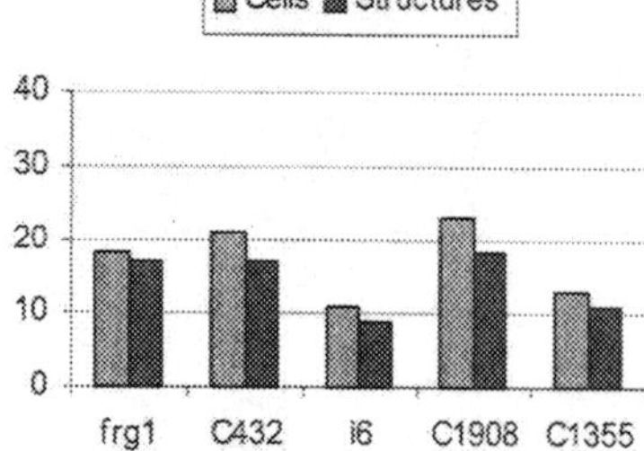


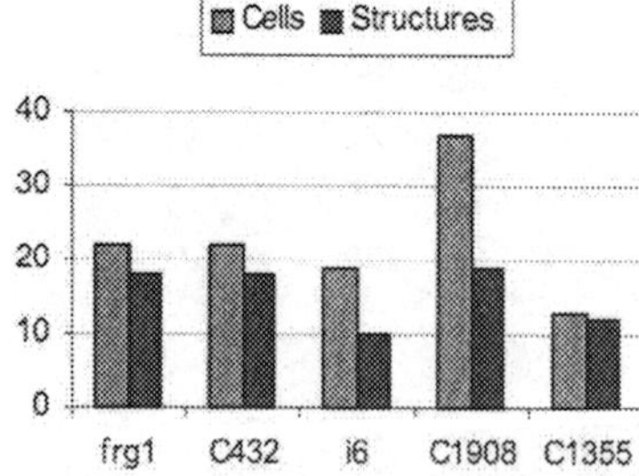

**FIGURE 16.** Number of gates compared to unique circuit structures across varying benchmark sizes, synthesized using various libraries.

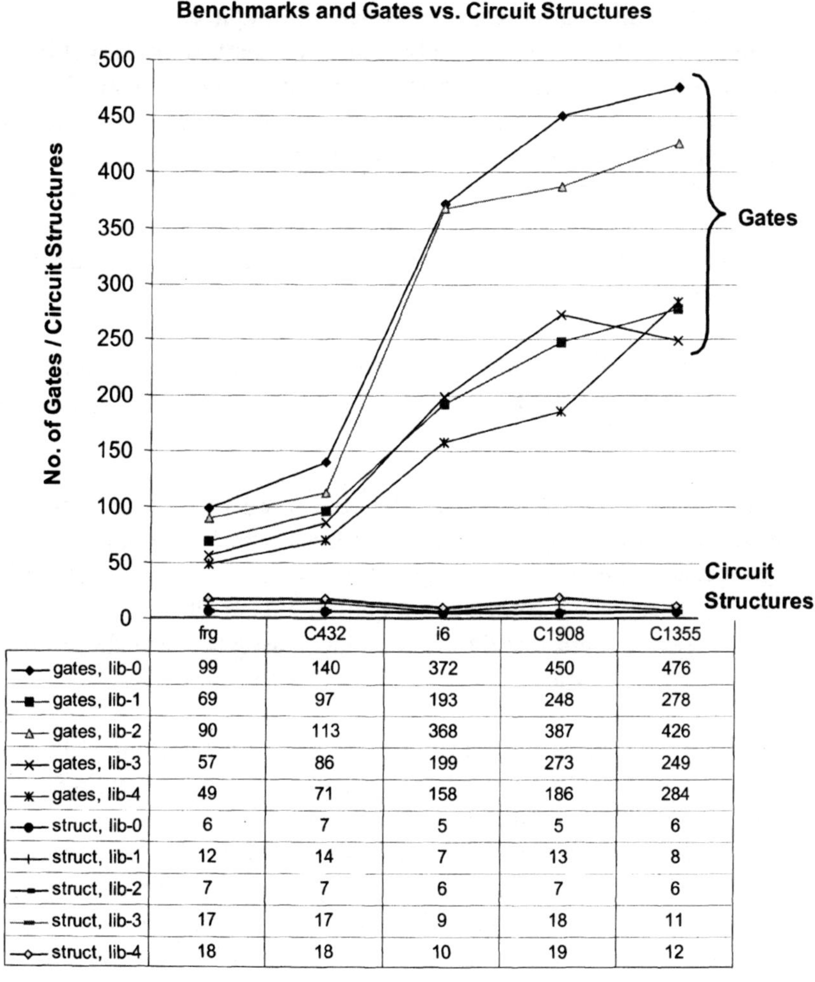



| | frg | C432 | i6 | C1908 | C1355 |
|---|---|---|---|---|---|
| gates, lib-0 | 99 | 140 | 372 | 450 | 476 |
| gates, lib-1 | 69 | 97 | 193 | 248 | 278 |
| gates, lib-2 | 90 | 113 | 368 | 387 | 426 |
| gates, lib-3 | 57 | 86 | 199 | 273 | 249 |
| gates, lib-4 | 49 | 71 | 158 | 186 | 284 |
| struct, lib-0 | 6 | 7 | 5 | 5 | 6 |
| struct, lib-1 | 12 | 14 | 7 | 13 | 8 |
| struct, lib-2 | 7 | 7 | 6 | 7 | 6 |
| struct, lib-3 | 17 | 17 | 9 | 18 | 11 |
| struct, lib-4 | 18 | 18 | 10 | 19 | 12 |

For the transistor-level netlists that resulted from the synthesis experiments, we now look at the circuit structures identified using the techniques described in this chapter. For each target library, Figure 15 compares the number of unique cells from the library that got used in the various benchmarks with the number of unique circuit structures in the netlist. As before, these numbers are a property of the target library used. For the simple library (lib-0), the number of cells equals the number of circuit structures. As the library size increases, complex gates like ANDs & AOIs break up into multiple individual circuit structures. However, different cells of varying sizes could map to the same circuit structure. Notice that in our experiments, as the library size increased, the number of unique circuit structures were smaller than the number of unique cells from the library that got used.

This is significant because we just showed that these numbers are far fewer than the total number of elements in the library, for large libraries.

In Figure 16, we have sorted the benchmarks based on their sizes. Notice that as the benchmark size increases, while the total number of logic gates in the netlist rises (and consequently the number of transistors increases), the number of unique circuit structures in the netlist does not scale significantly. This is the reason we use these circuit structures as the basis of our divide-and-conquer strategy.

The trends we observe in our experiments are in tune with other published work related to synthesis and technology mapping. Early work comparing transistor implementations of logic cells and trade-offs on the cell library composition was done by Rudell [Rudell 89]. It was shown in [Detjens 98] that significant improvement can be achieved by having a larger, more complex library, where the number of series and parallel transistors, in the static CMOS cells, is varied. [Guan 96] later showed that even though the total number of such complex cells can get to the order of billions, the technology mapper only selected on the

**FIGURE 17.** Circuit Structure Library for circuit C432

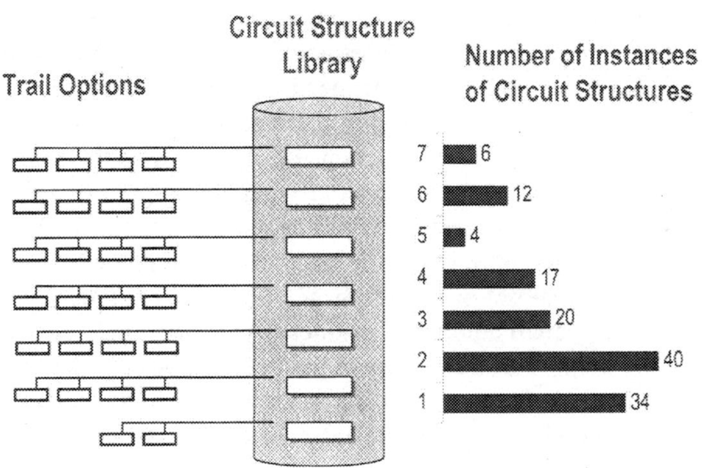


order of 100 of these cells to implement each of their logic synthesis benchmark circuits.

## 2.6.3 Circuit Structure Library Example

For one of the benchmarks from the experiments above, namely C432 with 113 gates synthesized using lib-2 above, the resulting transistor-level netlist had 578 transistors and 133 clusters but only 7 distinct circuit structures. The resulting circuit structure library is illustrated in Figure 17, showing the number of instances of each of those circuit structures in the netlist and also the number of trail pattern options for each of them. Notice that each of the elements has a very small number of trail pattern options to choose from. This is the reason for reduced computational effort during layout generation.

## 2.7 Summary

In this chapter, we discussed the importance of trails in shape-level layout and showed how groups of strongly connected transistors and their trail formations can be abstracted as clusters. We showed the necessity for clustering and the need to keep it minimal. We then described strategies for identifying these clusters in a netlist and using pattern matching to recognize groups of clusters with the same circuit structure. Such a strategy helps us efficiently compute trail formation options. The results presented in this chapter demonstrate that while the libraries used for synthesis may be large, the resulting netlists make use of a very small subset of cells from the library. Consequently, we end up with a very small number of distinct circuit structures. Further, even though the number of transistors in the netlists goes up, the number of distinct circuit structures does not scale significantly, justifying our choice of circuit structures as the basis for divide-and-conquer.

In the next few chapters, we describe how these essential clusters and their trail formation options form the basis of our placement algorithms. We first place these clusters globally while minimizing wirelength, congestion and timing. These clusters however remain uncommitted to any shape-level layout until our detailed placement step, where we explore various shape-level layout options, in the context of local optimality.

     *Global Placement*

## 3.1 Introduction

While clustering forms the essence of our divide and conquer approach, the novel idea in our placement technique is to find an approximate placement for these clusters before committing each cluster to a specific shape-level layout. Such a two phase approach is appropriate because global aspects of the placement like wirelength, congestion, area, row-densities and timing, depend on the overall locations of these clusters, while the detailed shape-level optimizations are affected by inter-cluster interactions in the context of local optimality. In this chapter, we focus on our techniques for global placement of uncommitted clusters, while optimizing for wirelength, congestion, area and row-densities. In a later chapter, we explain how this method is extended to handle timing optimization.

Placement is a well researched topic. Macro-placement methods like quadratic placement, combined with min-cut partitioning strategies ([Tsay 88], [Kleinhans 91], [Eisenmann 98]), have proven to be very effective in reducing overall congestion. On the other hand, iterative

**FIGURE 18.** Global Placement Strategy

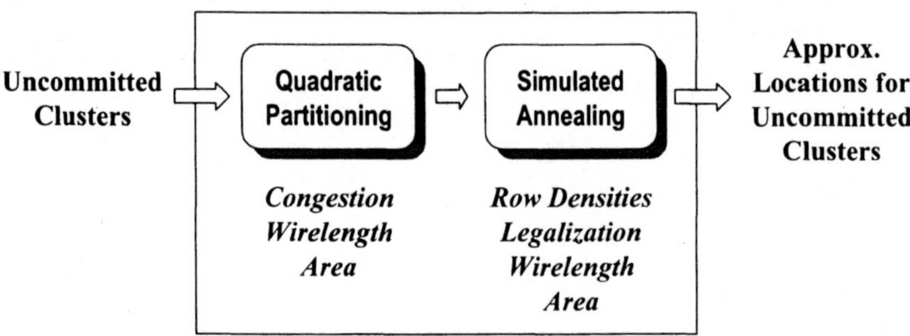



improvement techniques that use simulated annealing ([Sechen 86]) have useful hill-climbing abilities that allow for multi-objective optimization. While our techniques are not limited by the choice of a placement algorithm, we have chosen a combination of some of these standard techniques to explicitly address our placement objectives. As shown in Figure 18, we use a combination of quadratic placement and partitioning techniques to quickly optimize for wirelength, congestion & area. This is followed by simulated annealing based iterative improvement to optimize for row-densities and legal non-overlapping cluster placement, while also accounting for area and wirelength. In the rest of this chapter, we describe these techniques in detail and also present some results to demonstrate their utility. During this phase, clusters, that are the placeable objects, are modeled using an approximation for their widths & heights. This approximation does not affect the final result significantly, since the next placement phase (detailed placement) handles shape-level optimization. Also, as a convenient simplification, we assume the input/output pins are located along the boundary of the block.

**FIGURE 19.** Quadratic Partitioning

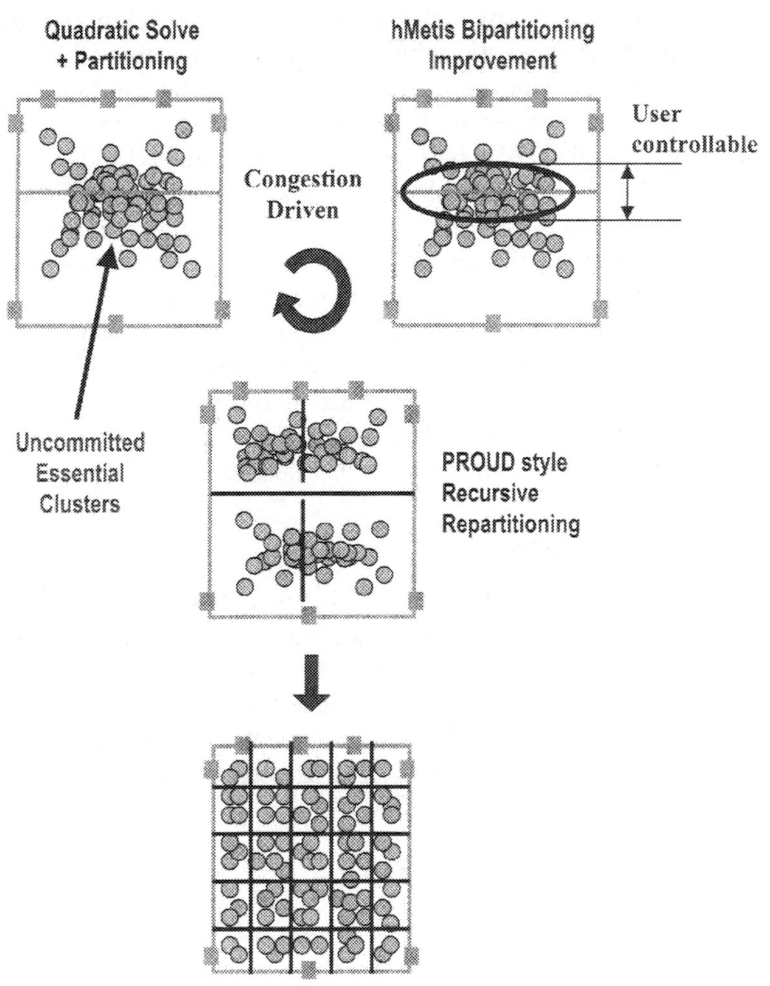

## 3.2 Quadratic Placement and Partitioning

We employ recursive re-partitioning-based quadratic placement, in the style of PROUD [Tsay 88]. This involves three major components: (a) quadratic solve, (b) recursive re-partitioning, (c) bi-partitioning improvement. A quadratic solve gives us initial locations for the clusters. This is used as a starting point for partitioning. During the partitioning phase, a horizontal (or vertical) cut line is located, via sorting on Y (or X), so that about half the cluster area is on each side of the cut. A physical cutline is then placed at the actual center of the partition. For each partition thus formed, quadratic optimization is carried on, for clusters within the partition, after projecting the coordinates of all clusters/pins outside the partition onto the cutline. Applying this idea recursively (see Figure 19), we create a sequence of increasingly smaller (but increasing in number), wide and narrow regions in which we confine and re-place the enclosed modules. In this section, we briefly look at some of the highlights of these steps. The reader is recommended to refer to relevant publications for further details.

### 3.2.1 Quadratic Solve

As in [Tsay 88] & [Kleinhans 91], a quadratic programming problem is derived from the circuit connectivity and is solved to generate an initial location for each of the clusters. During this solve, the clusters are modeled as point-modules and the nets are modeled as two-point connections between all the modules connected to them. The objective function is the weighted sum of the squared rubber-band lengths of nets:

$$\Phi(x, y) = \frac{1}{2}x^T Cx + d_x^T x + \frac{1}{2}y^T Cy + d_y^T y$$

**FIGURE 20.** Setting up objective function for quadratic solve

```
procedure set_up_objective_function
    Initialize C = 0, dx = 0, dy = 0                    16
    foreach net in Netlist, n                          17
        wt = net_wt(n) * 2 / (num_pads(n) +
        num_modules(n))                                18
        foreach module connected to net n, μn          19
            C[μn , μn] += wt * (num_pads(n) +
            num_modules(n) - 1)                        20
            foreach pad connected to net n, pn21
                dx[μn] += wt *
                pad_x_location(pn)                     22
                dy[μn] += wt *
                pad_y_location(pn)                     23
            end                                        24
        end                                            25
    end                                                26
end                                                    27
```

The vectors $x$ and $y$ represent the coordinates of the various clusters for the movable modules. The matrix $C$, also referred to as the *modified connectivity matrix*, and the vectors $d_x$ & $d_y$ are populated by the procedure *set_up_objective_function*, described in Figure 20. Here, *num_mods(n)* refers to the number of movable modules connected to net *n* and *num_pads(n)* refers to the number of pads connected to net *n*. The pads here correspond to either the input-output pins along the boundaries, or the modules in other partitions that are considered fixed for the purposes of the solve within a given partition. *pad_x_location* and *pad_y_location* refer to the locations of the pads projected onto the boundary of the current partition. The parameter *net_wt* is an optional multiplier to scale the net's importance. This is used, later on, for timing-driven placement. The x and y optimizations are separable and the solution is found by solving the following set of linear equations:

$$Cx + d_x = 0$$

$$Cy + d_y = 0$$

We solved these equations using conjugate-gradient based methods provided in the LASPack library [Skalicky 96].

### 3.2.2 Recursive Re-Partitioning

In the style of PROUD, the quadratic solve & partitioning techniques are applied recursively to the resulting smaller partitions. In order to capture the effect of modules in other partitions to the placement of modules in the current partition, we resort to terminal propagation, where modules & pins outside the current partition are treated like pads and their locations are projected to the boundaries of the current partition as illustrated in Figure 21.

### 3.2.3 Bi-Partitioning Improvement

Quadratic solves invariably result in many modules placed very close to the cutline, making it difficult for the partitioning step to make intelligent decisions. GORDIAN [Kleinhans 91] introduced the idea of using bipartitioning to resolve which side of the cut these objects

**FIGURE 21.** PROUD-style terminal propagation during recursive re-partitioning

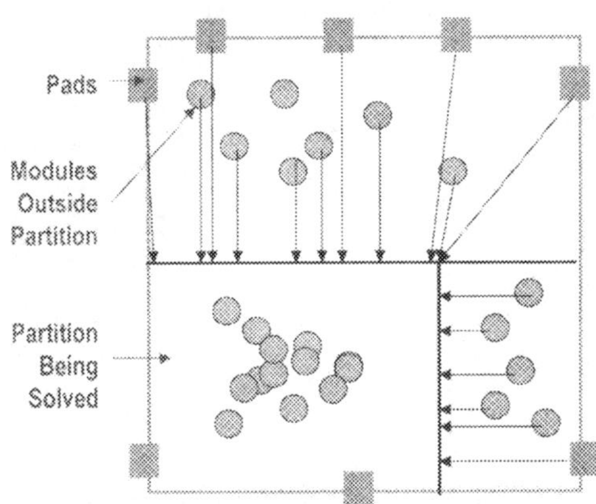


should be bound to. Hence, we formulate a similar bipartitioning problem that is allowed to relocate a specified fraction of the clusters on each side of the physical partition, that are near the cutline. This is also shown in Figure 19. The remaining clusters are assigned to partitions based on their quadratic solve-based locations. Varying this fraction lets us control the emphasis on bipartitioning relative to that on pure quadratic placement. As suggested in [Kleinhans 91], a fraction of about 0.5 gives us good results on average. For bipartitioning, we used hMetis [Karypis 97].

## 3.3 Simulated Annealing Legalization

The final step during global placement is a phase of simulated annealing ([Kirkpatrick 83], [Sechen 85], [Sechen 86], [Rutenbar 89]). The result of quadratic partitioning does not necessarily place all mod-

**FIGURE 22.** Simulated Annealing legalization

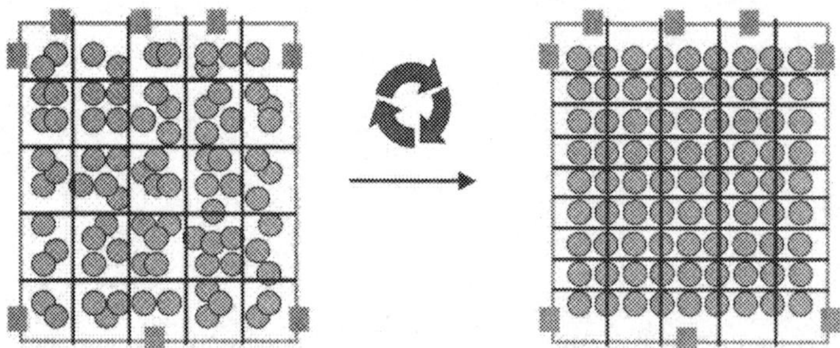


ules in rows or in a non-overlapping fashion. Iterative improvement is therefore necessary for legalization, local minimization of wirelength and distributing modules appropriately amongst the various rows, as illustrated in Figure 22.

The following are the four major components of our annealing formulation:

- **State Representation**: The layout area is divided into a grid where the horizontal grid-lines correspond to the various rows in the layout. The vertical grid-lines are drawn such that a particular grid-location can hold a few modules. The distribution of the modules amongst the various grid-locations defines a state in our optimization.

- **Move Set**: This comprises random re-allocations of modules between the various grid-locations. It includes both swapping of module locations and moving modules to other grid locations. To keep the optimization local, the moves are range-limited to a few grid-locations.

- **Cost Function**: There are three essential terms in our cost function.

  *Cost = w1 . NetCost + w2 . GridCost + w3 . RowCost*

  *NetCost* is evaluated as the sum of the half-perimeters of the net-bounding boxes for all nets. This optimizes for wirelength.

  *GridCost* is used to legalize overlaps between the various modules. It is a ratio of the module widths at a particular grid-location to the capacity of the grid-location.

  RowCost is used to optimize for row-raggedness and also row-utilization. It is a measure of the variances between various row widths. This helps minimize the area of the layout. It also measures the ratio of row utilization to the targeted row-utilization, where utilization is a measure of the sum of the widths of the all modules in a row. As we show later in our results, row-utilization can also be exploited to control congestion in the layout.

- **Cooling Schedule**: We use a Modified Lam cooling schedule [Lam 88] [Swartz 90] that allows us to set preferred acceptance ratios at the various temperature regimes of the annealer. In order to keep the resulting placement changes minimal, we perform annealing in the cold regime, using a very low initial temperature. This favors down-hill optimization.

We used the Anneal++ library [Krasnicki 97], which provides a rich set of schedule optimizations and also mechanisms that dynamically modulate the probabilities for when each of the annealing moves is selected, using the Hustin scheme [Hustin 87], as annealing proceeds.

In the next section, we present results from our experiments to demonstrate the effectiveness of each of these techniques in optimizing for the relevant placement goals.

## 3.4 Global Placement Results

For the global placement experiments (and detailed placement experiments in next chapter), we used parameters from the HP technology and the LGSynth91 benchmarks as described in the previous chapter. While the results comparing fully routed results are presented in the

**FIGURE 23.** Congestion plots for C432 during global placement

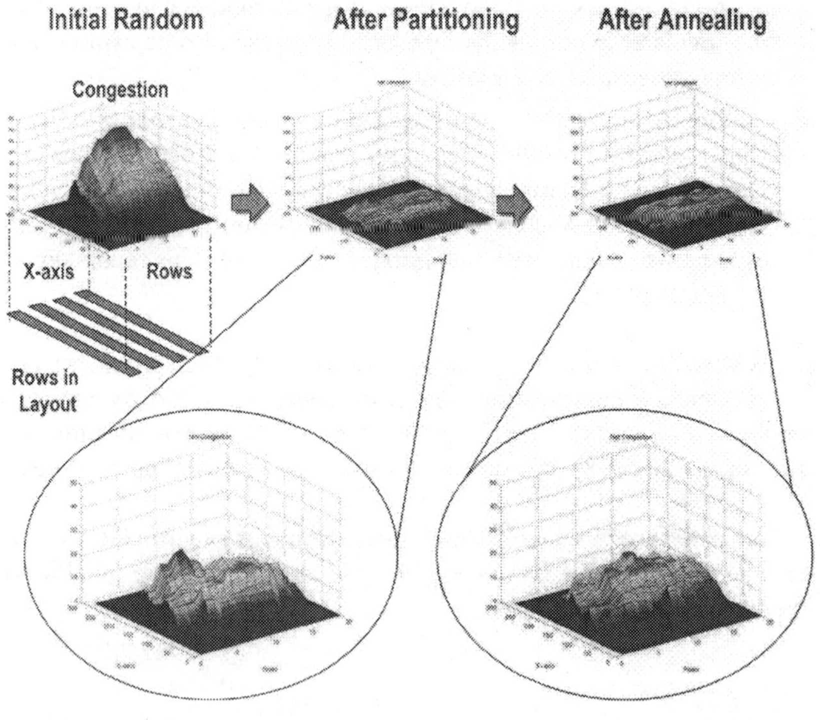

next chapter, we present here some insights into how congestion is minimized during the various stages of global placement. We also demonstrate how congestion can be manipulated during this phase.

### 3.4.1 Congestion during Global Placement

In Figure 23, we present congestion diagrams for the circuit C432, with 836 transistors and 456 nets. The congestion at a given location in the layout is measured as the number of bounding boxes of nets crossing that particular location. Notice that quadratic partitioning is very effective at reducing the overall congestion in the layout significantly. It also runs in a much smaller time. Notice also that our annealing step maintains the global congestion at approximately the same level, even improving it a little bit, while making small perturbations to the layout to redistribute the modules.

### 3.4.2 Row Utilization and Congestion

As mentioned earlier, row utilization, defined as the total width of modules in the row, can be used to control congestion in the layout. Some large circuits, like C1908 (with 2552 transistors and 1311 nets), result in very congested layouts, when placed as tight as possible. This adversely affects their routability. We can relieve this congestion by reducing the module densities in the various rows, thereby compromising on the area of the layout. Further, such reduced row densities can be varied differently for the various rows. In our experience, rows in the middle tend to get more congested than rows towards the top or bottom of the layout. The densities of the modules in the rows, can be controlled during the annealing phase through the *rowCost* cost component. We demonstrate this with an example.

**FIGURE 24.** Effect of row utilization on congestion for circuit C1908

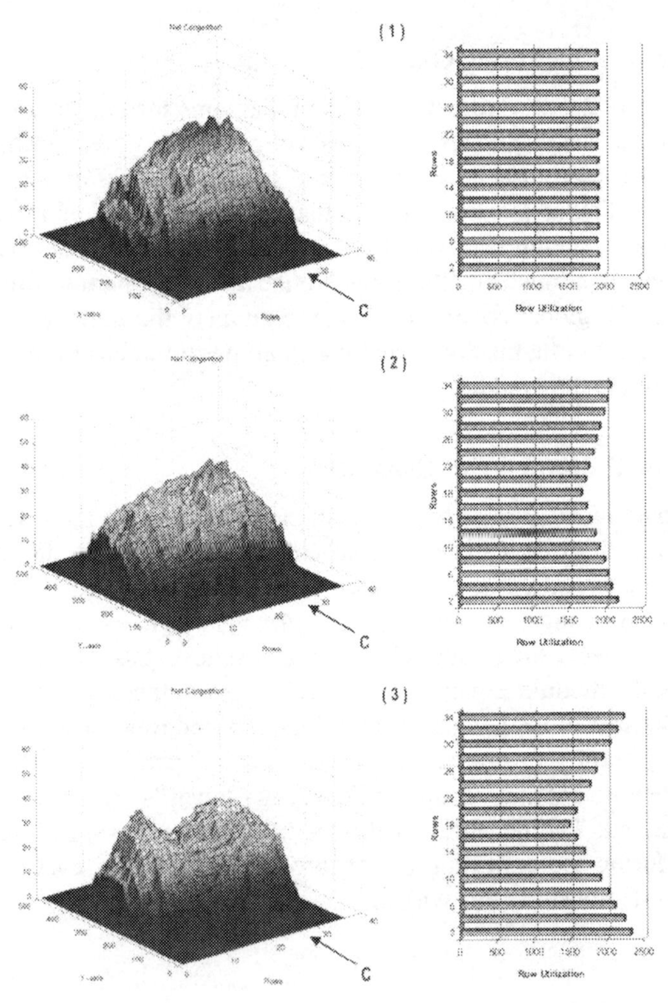

In Figure 24, we show the resulting congestion plots for an example circuit, C1908, for 3 cases: (1) equal minimum row utilization (2) small variation in row utilization (3) large variation in row utilization. For the row utilization variations, we have used a linear gradation as shown in the figure, with minimum utilization at the center of the layout. Notice that by introducing the row variation, we are able to reduce the overall congestion for (2). In order to clearly see this difference, we compare congestions for the three cases, at a cross-section C shown in the figure, separately in Figure 25. This figure clearly shows the trade-off between reduced congestion and increased row width.

Notice also in Figure 24 that as the variation is increased further, like in (3), while congestion reduces in the middle, it starts increasing at other locations near the top or bottom of the layout. The optimal value by which to vary the utilization is benchmark dependent and we deduce it by trying out a few different values.

**FIGURE 25.** Congestion reduction and row width increase for C1908, at cross-section C, by varying row utilization

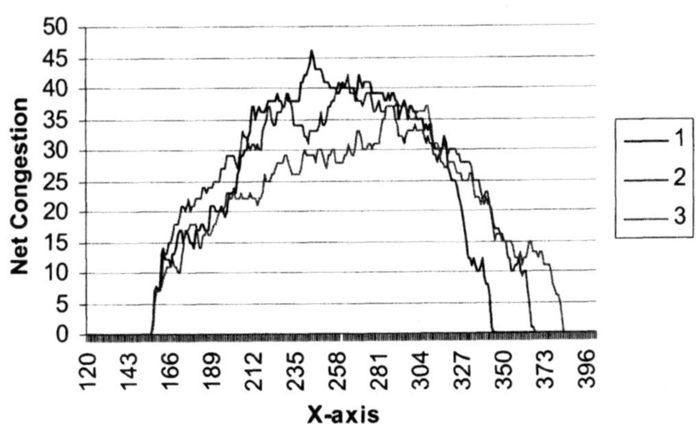

## 3.5 Summary

We described in this chapter how wirelength, congestion, area and row-densities, which are block level concerns, are addressed by our first phase of placement. In a later chapter, we describe how timing optimization is incorporated into this phase. Most prior transistor-level placers and essentially all standard-cell placers would stop here. These approaches do not permit *inter-cluster merging*. This is primarily because denser diffusion merges usually compromise routability. At block-level, standard cell designs are routable at least in part because at transistor level, all the internal sub-nets are fully pre-routed and sufficient pin spacings have been arranged for net accessibility. Because we do not wish to stop here, however, we must take the responsibility for these careful shape-level optimizations. This is the purpose of the next phase of placement, discussed in the next chapter.

# CHAPTER 4

# Detailed Placement And Layout Results

## 4.1 Introduction

At the end of global placement, we have arranged device clusters in rows, legalized these clusters, and minimized an estimate of wirelength and congestion. However, none of these clusters have yet been committed to a physical trail-level layout. By deferring this binding to a separate placement step, we are able to optimize each cluster carefully against a good global model of both wirelength and congestion.

In this chapter, we discuss the second phase of placement, where we optimize individual clusters and the boundaries between clusters for density, and for routability. Attention to shape-level routability turns out to be extremely crucial here. We use simple global routing to predict macroscopic routability during this phase. We then present a novel technique to generate dense layouts while handling local and global routability considerations. This chapter also describes our routing integration that creates fully routed layout results. We conclude by presenting layout results from experiments comparing our flow to a standard cell-based flow.

**FIGURE 26.** Detailed Placement Strategy

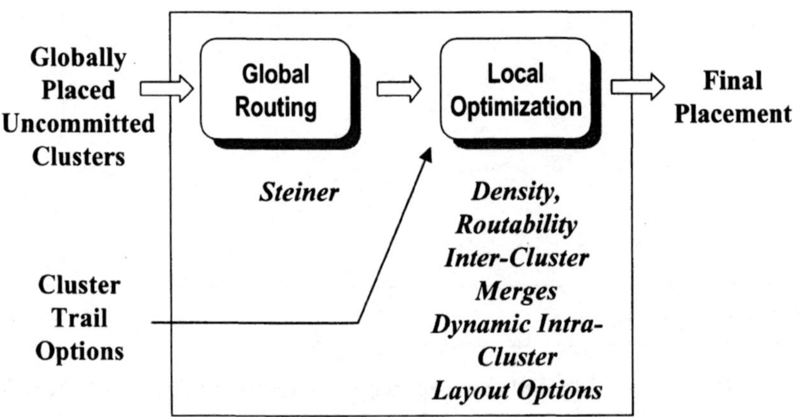



Figure 26 gives an overview of our strategy. We start with a global routing step that is based on Steiner routing. This is then used to model global congestion during our local placement optimization and is also dynamically updated in the process. We first discuss how density and routability (both local & global) are addressed during intra-cluster and inter-cluster interactions. We then describe how we optimize for them during our local placement optimization.

## 4.2 Intra-Cluster Optimizations

Our essential clusters are typically composed of a few transistor trails. The cluster is "uncommitted"—it has a nominal width in the row, but no specific layout. To create a palette of layout alternatives for a cluster, we must focus carefully on routability of individual trails, and groups of adjacent trails.

To begin, note that a transistor trail is fully routable *only if*:

- All sub-nets internal to the trail have enough space to be completely routed within the trail.
- All nets connecting to the outside have enough space to exit from the trail through pin escapes.

Figure 27 illustrates this for a simple example. It is important to ensure that pins connecting to polysilicon gate inputs are spaced appropriately to allow accessibility on metal layers.

**FIGURE 27.** Trail Routability

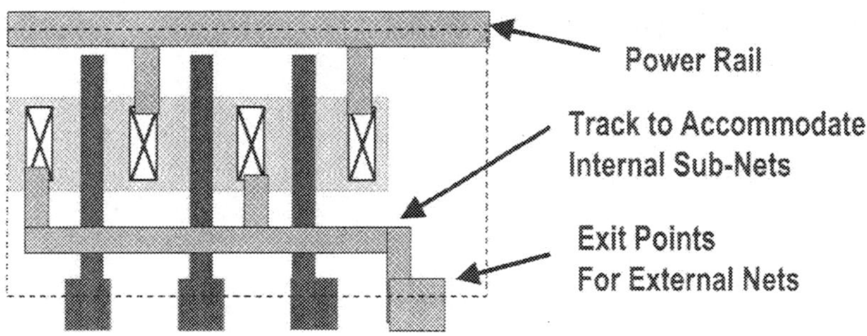

The same conditions apply when we need to route a set of trails. For example, a pair of vertically adjacent trails in a cluster typically represent NMOS and PMOS devices sharing some poly gates; choosing the placement so that poly gates can both align and escape to metal is critical. A pair of horizontally adjacent trails typically represent same-type MOS devices that may be able to share diffusion and merge. Again, the critical routability issue is ensuring enough tracks so that internal subnets can connect to these devices, while not compromising any polysilicon gate alignments.

Our first step in generating cluster layout options is to match the cluster circuit structure with patterns in the circuit structure library. As described earlier, this gives us the various feasible combinations of transistors to produce easily routable trails. Each of those trail formations then offers us layout alternatives for the cluster. We note here that while some small gates (like NAND, NOR) map one-to-one to essential clusters, other small gates (like AND, OR) and larger gates (like AOI, OAI) map to a handful of essential clusters. Relative geometric locations are then assigned to these trails while respecting shape-level routability issues we just discussed. This is then combined with transistor-size information to create a shape-level binding (see Figure 28). For big transistors, various fingering options can be incorporated. This can result in a richer set of cluster layout options, thereby supporting dynamic folding during local placement optimization.

**FIGURE 28.** Layout formation options and shape-level binding

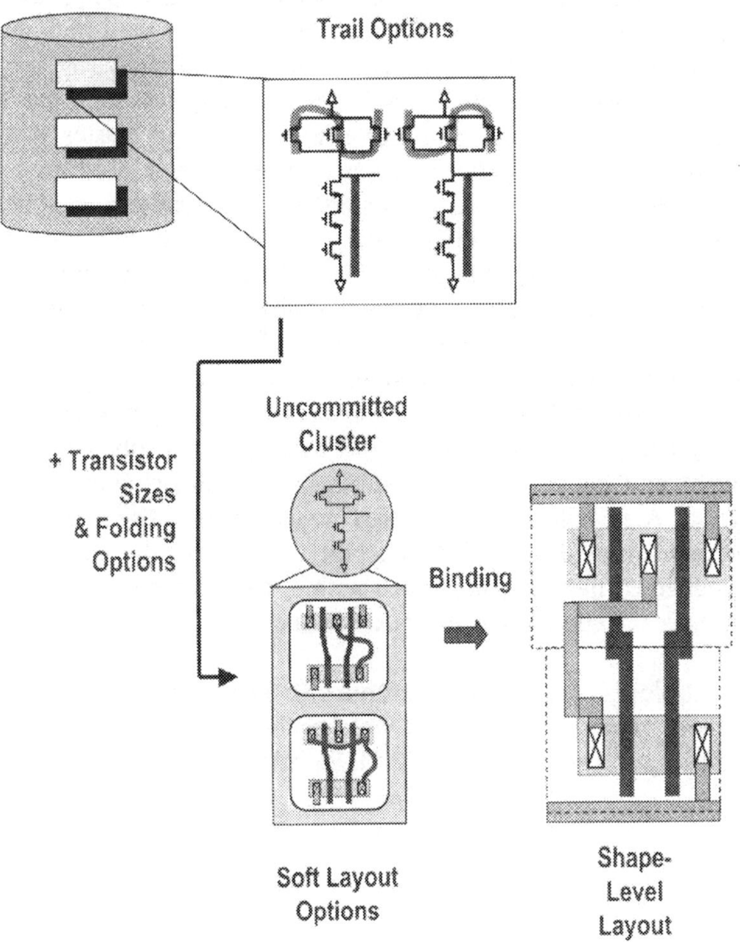

## 4.3 Inter-Cluster Optimizations

Allowing adjacent clusters to merge, i.e., to share diffusion in the N- or P-type device rows, is a simple but remarkably powerful optimization. In particular, it is a local optimization that is *amplified* when we scale to thousands of devices. We illustrate this with a simple example. Figure 29 (left) shows the different intra-cluster layout options for a simple circuit cluster. Given a placed cluster, we show (on the right) different possibilities of inter-cluster merging between the layout options for the uncommitted cluster and the placed cluster. Notice in the figure that if neighboring clusters cannot merge (i.e., no suitable selection/orientation of the uncommitted cluster yields a legal geometric merge) then the resulting layout resembles a pair of standard cells: the clusters can only abut at their boundaries. But if a suitable merge can be found, the clusters can actually *overlap*. In this simple example, the best merging option allows diffusion sharing in the top device row (and slightly extends the diffusions if necessary for design rules), and replaces one small power rail connection with a single wider strap that is physically on the boundary between the clusters. Notice the saving in area, even for a single inter-cluster merge, by choosing the *best* layout option. Also note that all merges we consider between neighboring clusters satisfy the detailed routability conditions: the resulting dense layout must have enough space within, to accommodate all internal sub-nets and enough space for external nets to escape out.

At block-level, these layouts are also subject to global routing congestion. This is irrespective of whether the layout was generated using standard cells or at transistor-level. We address this next.

**FIGURE 29.** Best Option Merging

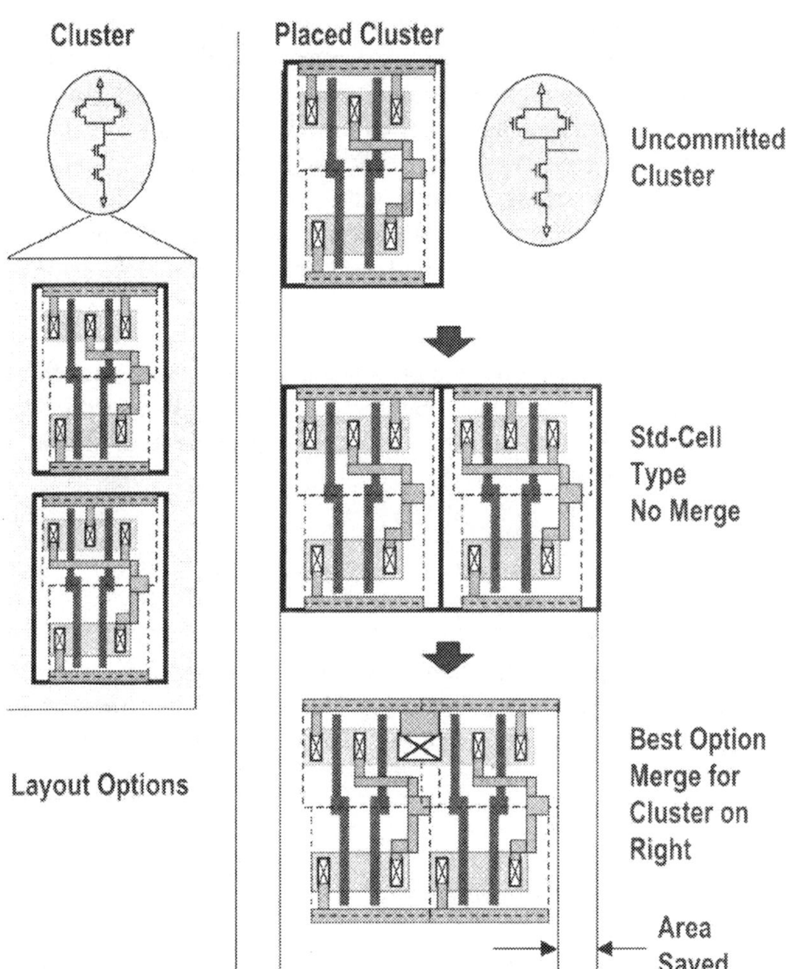

## 4.4 Global Routability

In order to estimate global routing congestion after global place-ment, we replace each net with its simple Steiner representation. For this purpose, we used Steiner algorithms from "The Generic C Library" [Madden]. We then calculate wiring demand at various locations in the layout using a coarse global routing grid. In regions that are heavily congested, we provide some relief by introducing extra vertical routing tracks, when we consider inter-cluster neighbor merges in that region. This makes sense because ultimately the pin locations of the trails con-strain global nets. The farther they are spaced, more the availability of routing tracks. Horizontal routing tracks are also introduced by adjust-ing row-spacings.

In the next sub-section, we present some linear-time algorithms to find the densest inter-cluster merges considering all possible layout options for each uncommitted cluster in the context of global and detailed routability.

## 4.5 Local Placement Optimization

In this section, we first present an analytical model for computing the best merge between neighboring clusters. We then present some row-based optimizations to find the best cluster layout option for each uncommitted cluster, while considering all merges between the various neighboring clusters.

### 4.5.1 Pair-Wise Cluster Merges

Consider two layout options L & R, as shown in Figure 30. Our goal is to compute the best merge, *BestMerge(L,R)*, that is the densest layout generated by merging layout option $R$ to the right of layout

**FIGURE 30.** Variable definitions during pair-wise cluster interactions

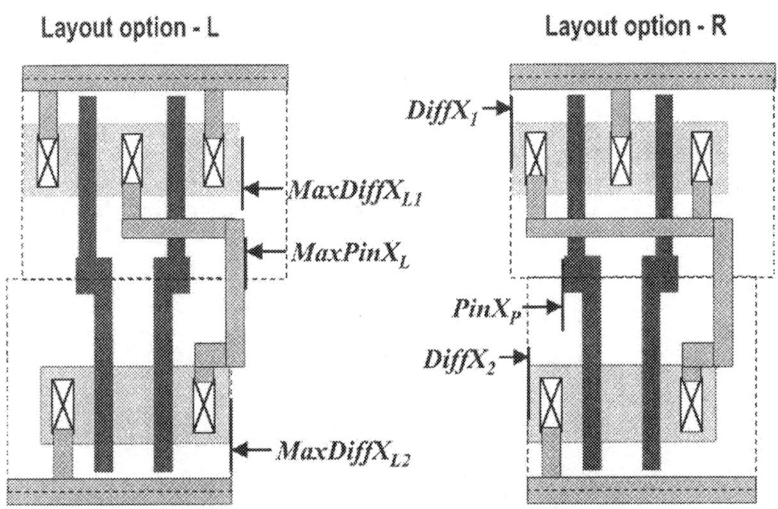


option $L$. Let $P$ be the set of all trails belonging to layout option $R$ that interact with $L$. Let $DiffX_p$ be the left-boundary of the diffusion region belonging to trail $p$ $(p \in P)$ and let $PinX_p$ be the location of the left-most pin belonging to $p$. Since the goal of BestMerge is to find the closest position layout option R can get to option L, we need to minimize $DiffX_p$ and $PinX_p$ over all trails $p$. This minimization is however subject to the following three sets of constraints: (a) Diffusion Merging, (b) Routability and (c) Poly Alignment.

## Diffusion Merging

The trails of the neighboring clusters can merge by sharing their diffusion regions. The distances between them are governed by the equation:

$$DiffX_p \geq MaxDiffX_{Lp} + DRCLength_p$$

**FIGURE 31.** Diffusion merging during pair-wise cluster interactions

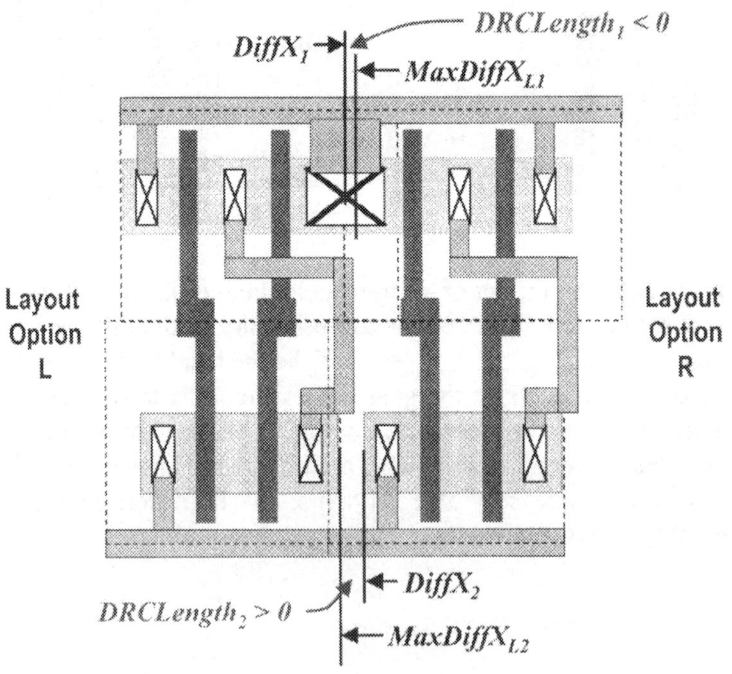

where $MaxDiffX_{Lp}$ is the right-most diffusion boundary of the trail belonging to layout option $L$ that interacts with $p$. $DRCLength_p$ is the design-rule imposed restriction. In general, if the interacting trails can be merged, then $DRCLength_p$ is negative, otherwise $DRCLength_p$ is positive. This is illustrated in Figure 31.

## Routability

We capture routability using a supply and demand model. A routing grid is overlaid on top of the layout area. Each grid location has a estimate of the number of horizontal and vertical tracks available for use. The Steiner routes are used compute the demand for global routing tracks in a given region. In heavily congested regions, the demand imposed by global routes is often larger than the supply of tracks. Under such circumstances, additional vertical tracks are introduced in between the clusters, in the hope that spacing them away from each other also increases spacing between the terminals connecting to those clusters, thereby relieving global congestion. In addition to the global routing demand, additional tracks may also be necessary to meet the local routing demands. We capture these by the equation:

$$PinX_p \geq MaxPinX_L + SubNetTracks + GlobalVertTracks$$

where $MaxPinX_L$ is the right-most location of a pin belonging to layout Option $L$ (see Figure 30). *SubNetTracks* are the number of extra vertical routing tracks required to completely route internal sub-nets belonging to the trail. *GlobalVertTracks* are the number of extra global vertical routing tracks required in this region.

## Poly Alignment

In order to align the polysilicon gate inputs of neighboring N & P trails and also the pins, we formulate a quadratic wirelength minimization equation between the objects that need alignment. Consider the example shown in Figure 32.

Variables $G_{pi}$ correspond to locations of the gates connected to net $i$ on trail $p$. Variables $P_i$ correspond to the locations of the pins. Using a

**FIGURE 32.** Poly gate alignment during pair-wise cluster interactions

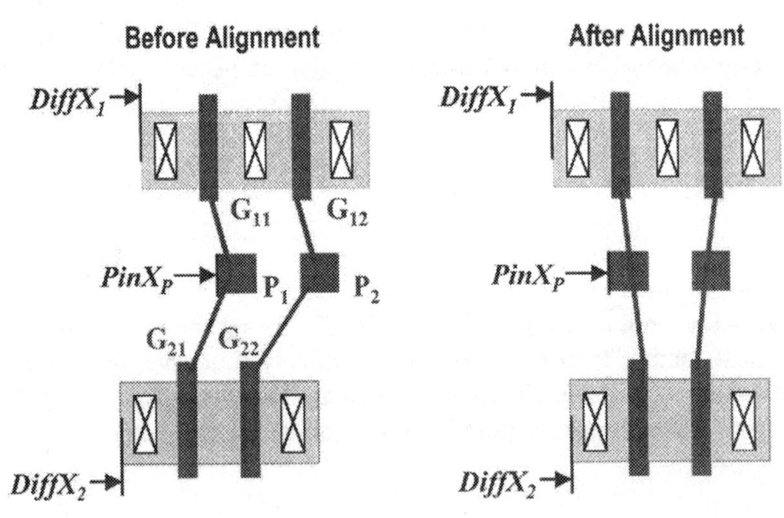

rubber-band model for the nets, the wirelength minimization objective is:

$$\textit{Minimize} \sum_{(p \,\in\, \textit{AlignP})} \sum_{(i \,\in\, \textit{GNets})} (G_{pi} - P_i)^2$$

where *AlignP* is the set of trails that needs to be aligned and *GNets* is the set of nets connected to the gates that need to be aligned. Assuming that the distances between the gate inputs are fixed, each $G_{pi}$ can be expressed as a sum of *DiffX$_p$* and a constant. Similarly, assuming that the distances between the pins is fixed, each $P_i$ can be expressed as a sum of *PinX$_p$* and a constant. Thus, the solution to such a quadratic formulation reduces to a linear constraint involving *DiffX$_p$* and *PinX$_p$*:

$$a_1 \times PinX_p + \sum_{p \,\in\, AlignP} (a_p \times DiffX_p) = const$$

where $a_1$, $a_p$ and *const* are scalar constants computed using the distances above. This corresponds to the illustration in Figure 32 (right).

To summarize, the goal of *BestMerge* is to then:

$$\binom{\textit{Minimize}}{\forall (p \in P)} \{DiffX_p, PinX_p\} \qquad subject \qquad to$$

- *Constraint A (Diffusion Merging)*

$$DiffX_p \geq MaxDiffX_L + DRCLength$$

- *Constraint B (Routability)*

$$PinX_p \geq MaxPinX_L + SubNetTracks + GlobalVertTracks$$

- *Constraint C (Poly Alignment)*

$$a_1 \times PinX_p + \sum_{p \in AlignP} (a_p \times DiffX_p) = const$$

*BestMerge* thus computes the left-most positions for all trails belonging to layout Option *R* while satisfying polysilicon alignment and routability constraints. Note that *Constraint B* dominates in regions that are global-routing congested, whereas *Constraint A* dominates in other regions, resulting in dense inter-cluster merges.

Overall layout optimization happens in a row-based fashion. We traverse the entire layout from bottom to top, while considering all neighboring cluster interactions. Within each row, we present two local optimization techniques for density and routability. These techniques make use of the BestMerge strategy between neighboring layout options. We now look at these two techniques.

**FIGURE 33.** Greedy Local Placement Optimization

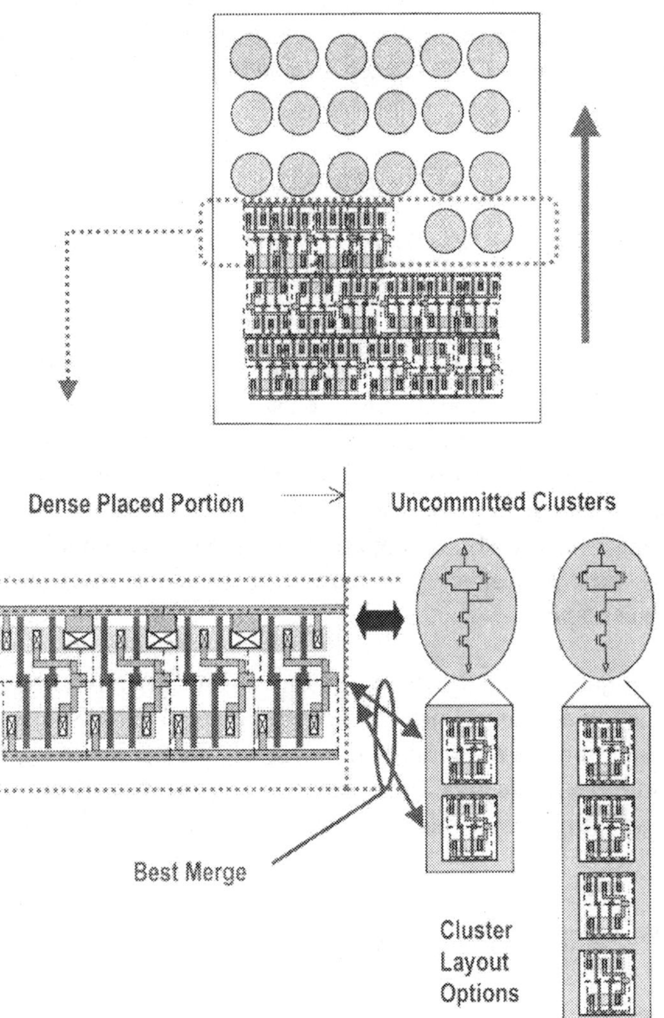

### 4.5.2 Greedy Local Optimization

We present here a greedy heuristic to compute the best layout option for each cluster within a row. As illustrated in Figure 33, we traverse each row from the left to right. For every cluster, we find the layout option that generates the best merge, given the best layout option for the neighboring (left) cluster. Formally, the best layout option for cluster $i+1$ is given by:

$$BestOption_{i+1} = k \qquad (k \in S_{i+1}) \qquad satisfying$$

$$MinWidth_{i+1} = MinWidth_i + BestMerge(BestOption_i, k)$$

where $S_i$ is the set of all layout options for cluster $i$, $MinWidth_i$ is the minimum row layout width up to cluster $i$. As mentioned above, the overall algorithm optimizes each row in the layout by traversing from bottom to top, left to right. At the end of processing every row, we update all global nets that were affected. This lets us use the most recent information while accounting for global congestion. During this process, inter-row spacing is introduced to satisfy horizontal global routing demand.

Even though greedy, this technique is very effective and fast. The implementation has a linear complexity of $O(K * n)$, where $K$ is the maximum number of layout options for any cluster in the circuit and $n$ is the number of clusters. In the next section, we present a more expensive, but optimal, dynamic programming approach, in the style of [Her 91].

### 4.5.3 Dynamic Programming Local Optimization

The greedy heuristic described above suffers from the drawback that early decisions made when traversing from left to right in the rows, can result in overall sub-optimal results. We can overcome this by formulating a dynamic programming problem, where the optimal layout width at any cluster, for a given layout option, can be computed by considering merges with all possible layout options at the previous cluster and their optimal layout widths (see Figure 34). In other words:

$$OptWidth(i+1, r) = \left( \underset{\forall (l \in L_i)}{Min} \right)(OptWidth(i, l) + BestMerge(l,r))$$

where *OptWidth(i,k)* refers to the optimal layout width at the $i^{th}$ cluster, when layout option *k* is selected. $L_i$ is the set of all layout options for the $i^{th}$ cluster. *BestMerge* is as defined previously. The optimal layout width for the row can then be computed as:

$$OptimalRowWidth = \left( \underset{\forall (l \in L_N)}{Min} \right)(OptWidth(N, l))$$

where *N* is the last cluster in the row. As with the greedy approach, the overall algorithm optimizes each row in the layout by traversing from bottom to top, while updating global routing and inter-row spacing appropriately. This approach scales linearly with the number of clusters as $O(K^2 * n)$, where *K* is the maximum number of layout options for any cluster in the circuit and *n* is the number of clusters. An obvious extension to both the greedy and dynamic programming approaches would be to iterate multiple passes through the complete row-based local optimization and global routing update.

**FIGURE 34.** Dynamic Programming Local Placement Optimization

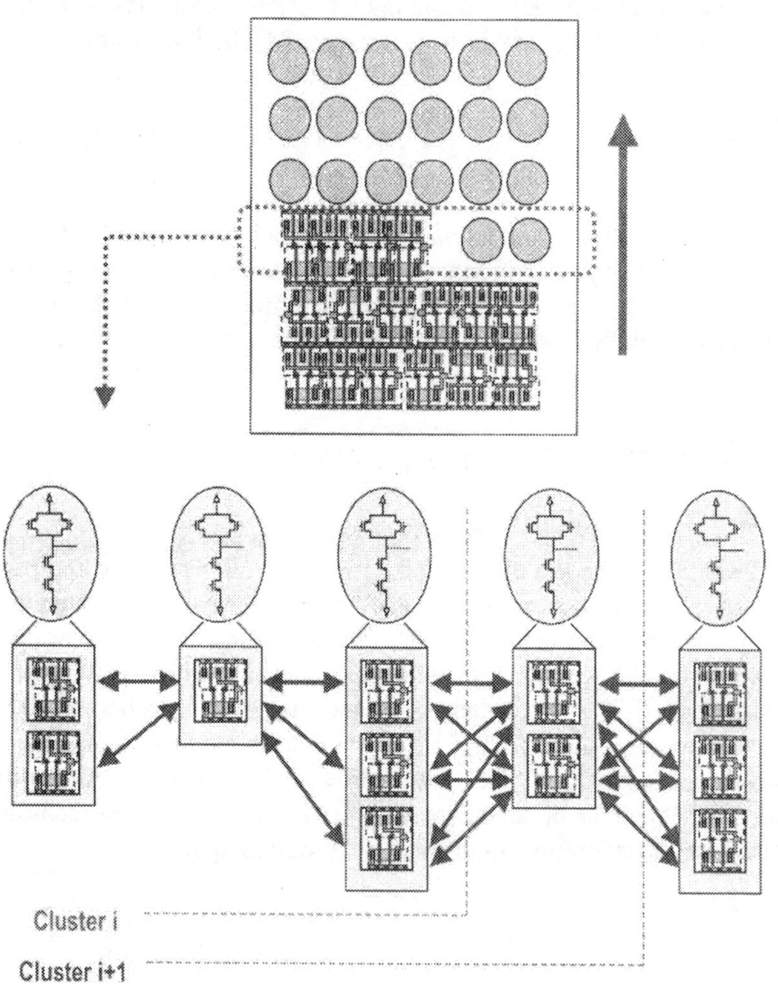

## 4.6 Routing Integration

To complete our flow, we have integrated the ICCraftsman router [Cadence] at the back-end. To enable routing completion, we constructively route the power nets, VDD & GND. We also place stubs of polysilicon corresponding to the intra-cluster gate pins. These pins are placed in a staggered fashion to enable easy polysilicon routing. We then invoke ICC to complete the local intra-cluster routing in Polysilicon and Metal1. ICC is invoked again to finish the inter-cluster routing using multiple Metal layers as required.

In the next section, we present our experimental setup and results comparing fully routed layouts generated using our tool with those generated using a standard Cadence cell-based flow.

## 4.7 Layout Results

Before we present our experimental setup and results, we take a quick look at a simple example result illustrating various algorithmic aspects of our flow.

### 4.7.1 Simple Layout Example

We describe here the results obtained for a simple benchmark, the circuit C432 from [CBL], synthesized using the HP technology. As described in Chapter 2, to create a suitable netlist for us, we re-synthesize C432 onto a simple target cell library that comprises INVERT, AND, NAND, OR, NOR gates up to 4 inputs. Assuming standard static CMOS, the circuit structure library has only 7 series-parallel circuit patterns in it, e.g., the AND gates end up decomposed into a combination of more simple NAND and INVERT patterns. Mapped onto this library, C432 has 191 gates; after flattening it has 836 devices and 456 nets; after essential cluster recognition it has 206 clusters which are

**FIGURE 35.** Congestion plots for circuit C432 after the two stages of

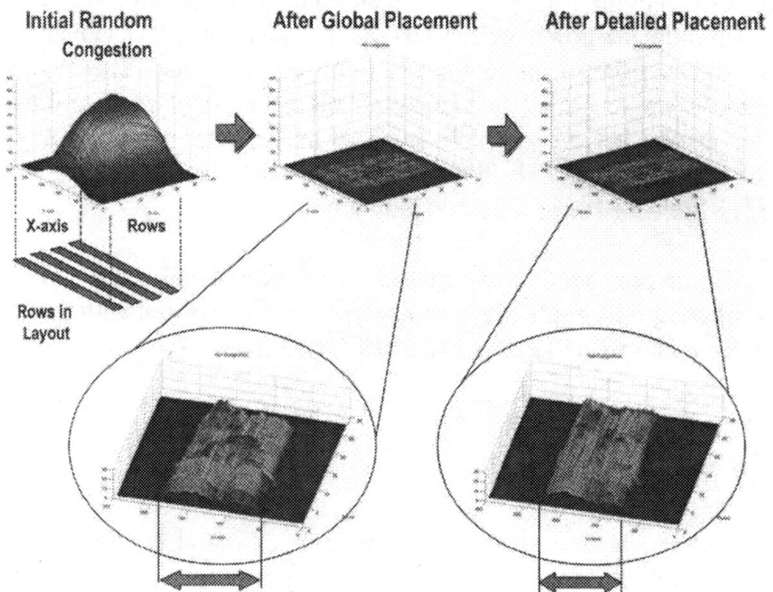


what we place in our global placer. The technology is 0.35um HP CMOS with one poly and 3 metal layers. Placement (global and detailed) requires 3 minutes; detail routing takes another 20 minutes.

Figure 35 shows congestion after both global and detailed placement, estimated at each point as the number of net bounding boxes that overlap that point. Notice that detailed placement reduces the width of the overall layout, without compromising overall congestion.

Figure 36 shows the distribution of trails of various sizes, where the size of a trail is the number of transistors in the trail. Detailed place-

**FIGURE 36.** Trail histograms for circuit C432

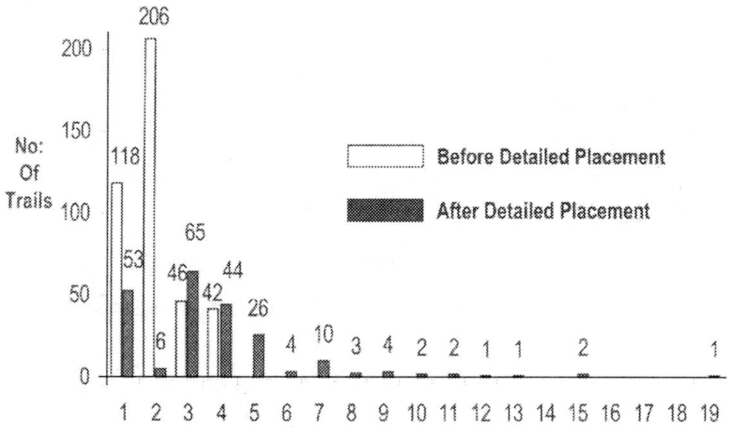


ment increases the number of "big" trails, i.e., it achieves significant diffusion merging across clusters.

Figure 37 displays a snapshot of TrailBlazer showing the final placement for C432 and the various stages of detailed routing. In the picture displaying poly and metal1 routing, notice dense diffusion merging, polysilicon gate alignment, staggered gate input pin locations, efficient metal1 usage and fully strapped diffusion terminals. In this picture, H is the same height as the standard cell row height. N/P refer to the NMOS/PMOS diffusion type. The fully routed layout has the same height as the standard cell-based layout but about 25% smaller area.

**FIGURE 37.** Placed and routed layout result for circuit C432

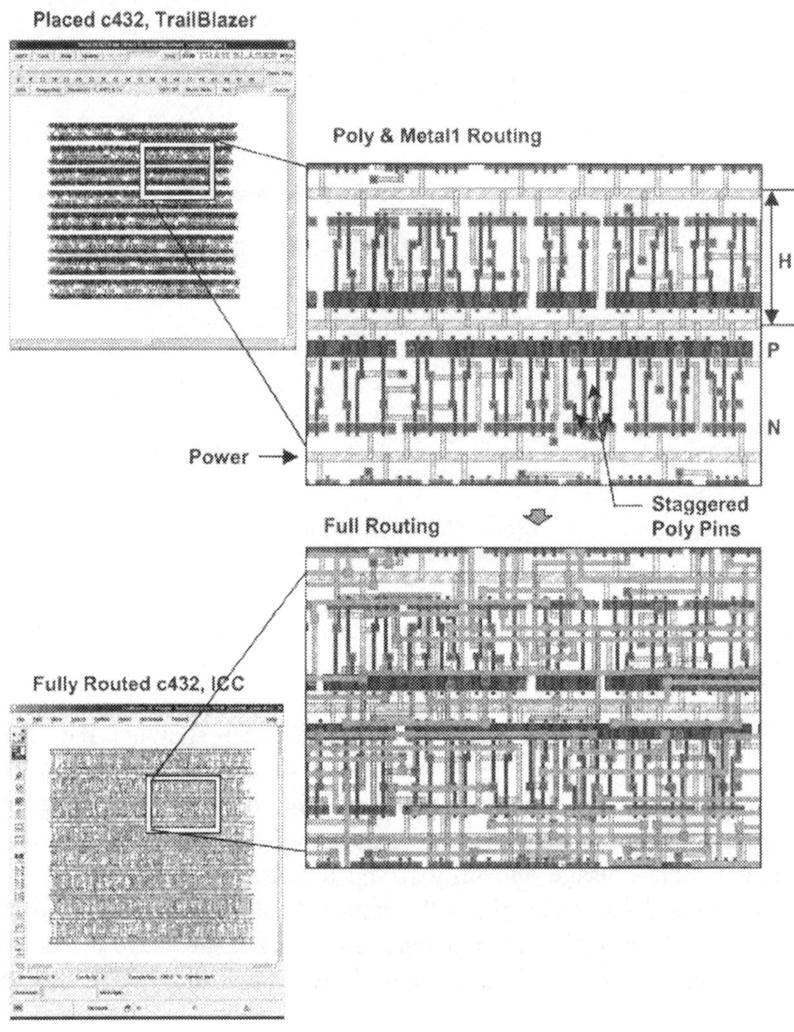

**FIGURE 38.** Experimental setup comparing layout areas using our flow with using a standard cell-based flow.

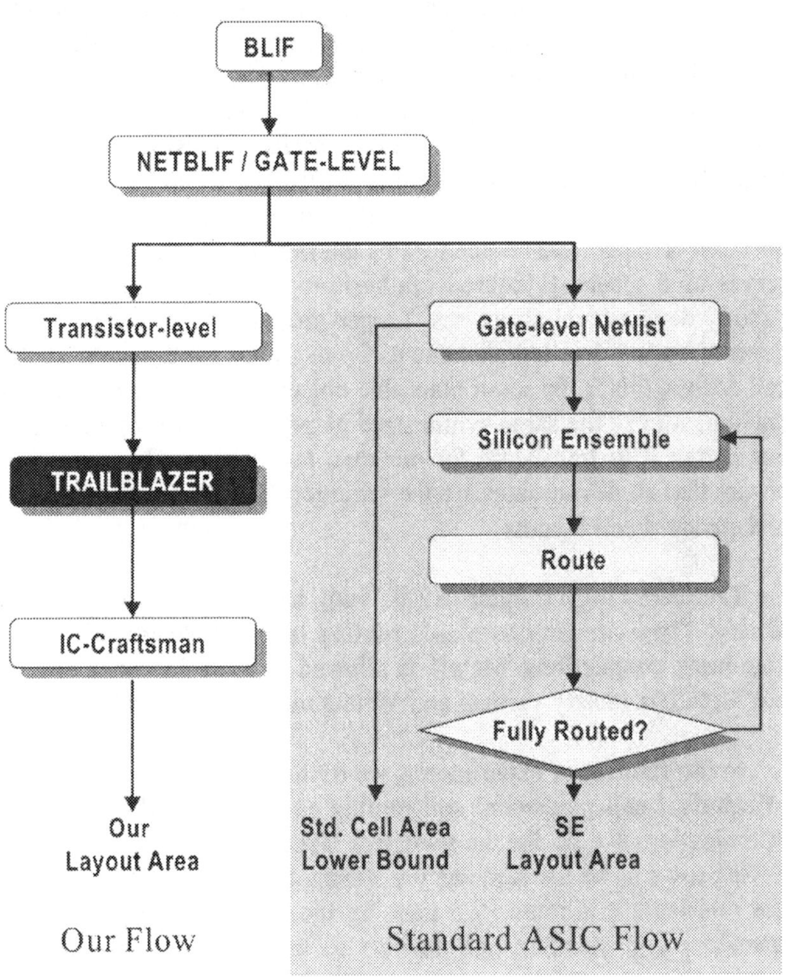

### 4.7.2 Layout Comparison Experiments

To quantify the potential advantages of transistor-based layout versus standard cells, we ran a series of logic netlists using our flow, and a standard Cadence Silicon Ensemble (SE) flow [Cadence]. As shown in Figure 38, and described in Chapter 2, we start with a logic description (in BLIF format) which is compiled into a functional schematic (in NETBLIF format) using logic synthesis (SIS [Sentovich 92]). We used several standard LGSynth91 benchmarks from [CBL]. The synthesis target library is again the simple INVERT/AND/NAND/OR/NOR library from the previous section. It is worth noting that a new and open question is the correct choice for this intermediate library: in our flow it serves as a means to coerce synthesis to produce a netlist that uses "good" device-level structures. Larger groupings for devices are discovered during detailed placement. On the other hand, for the standard cell design, this *is* the set of placeable objects. For the most direct comparison, we use the same synthesized gate-level netlist for each layout, but flatten it to transistors for our own layout flow. However, we do ensure that all device sizes are the same for the both the transistor-level and standard cell layouts.

The technology is again HP 0.35um, and we used an existing cell library. There are four available routing layers. Poly is only used for gate-input connections. Metal1 is allowed to route in either direction, but Metal2 is mostly vertical and Metal3 mostly horizontal.

In our first set of experiments, we try to normalize away the effects of standard cell placement and routing and focus on the lowest-level optimizations we do for the transistor layout. We fix the height of our device rows to be the same as the standard cells, and fix their pitch at the rule-legal minimum. We then fix the overall heights of both the transistor and standard cell layouts to be equal. We describe each benchmark, and compare total area of final routed layouts, in Table 1.4 . Note that we also compare against an *absolute lower bound*--the sum of the areas of the individual standard cells in the cell layout. Note that our layouts are 10-25% *better* than this absolute lower bound: this is because detailed placement allows clusters to overlap. We are also 19-29% better than the densest layouts generated using Silicon Ensemble (set to 99% row utilization, i.e., maximum placement density).

In our second set of experiments, we lay out a set of more difficult benchmarks that require more attention to routing congestion. To begin, we place our transistor-level netlists using TrailBlazer and then route them in Cadence. We then fix the height of the cell-based layout to be the same as our successful device-level layout. We adjusted percent row utilization in Silicon Ensemble until the cell layouts were just fully routable. Table 1.5 shows the results of this experiment. We note again that the transistor-level layouts are from 16% to 27% smaller than the comparable cell layout. The detailed placement optimization carefully

TABLE 1.4 Maximum Density Routed Results

| Bench. | No. of Trans | Rows | Our Layout Area (um$^2$) | Std. Cell Lower Bound (um$^2$) | %Impr over Lower Bound | Silicon Ensemble Area (um$^2$) | %Impr over SE |
|---|---|---|---|---|---|---|---|
| frg1 | 454 | 6 | 6714.9 | 8934.3 | 24.8% | 9452.7 | 28.9% |
| b9 | 464 | 5 | 7033.5 | 9153 | 23.12% | 9510.8 | 26% |
| i2 | 686 | 7 | 8968.1 | 11321.1 | 20.7% | 12606.3 | 28.8% |
| i4 | 764 | 7 | 11084.9 | 12360.6 | 10.3% | 13768.7 | 19.5% |
| C432 | 836 | 8 | 12744.0 | 16200.0 | 21.3% | 17172.0 | 25.7% |
| apex7 | 938 | 8 | 14850.0 | 18489.6 | 19.7% | 19828.8 | 25% |
| example2 | 1140 | 9 | 18893.3 | 22404.6 | 15.7% | 23874.8 | 20.9% |
| i6 | 2066 | 12 | 29224.8 | 34700.4 | 15.2% | 35964.0 | 18.8% |

**TABLE 1.5** Routing Congestion-Aware Routed Results

| Bench. | No. of Trans | Layout Height (um) | Our Layout Area $(um^2)$ | Silicon Ensemble Area $(um^2)$ | %Impr over SE |
|---|---|---|---|---|---|
| C880 | 1410 | 150 | 29595 | 38550 | 23.2% |
| term1 | 1682 | 162 | 34279 | 47142 | 27.3% |
| x4 | 1826 | 180 | 37620 | 46800 | 19.6 |
| C1355 | 2164 | 196 | 47844 | 61603 | 22.3% |
| C1908 | 2552 | 203 | 51887 | 61692 | 15.9% |
| i9 | 3256 | 244 | 67344 | 83204 | 19.1% |

allocates both intra-row wiring space and inter-row wiring space, and correlates quite well with the space required to actually complete the routing. We find empirically that with these routing optimizations disabled, it is impossible to create routable transistor-level layouts. The layouts were also generated in reasonable time. The largest benchmark with 3256 transistors took about 9 minutes to place and another 30 minutes to route on a desktop workstation.

### 4.7.3 Dynamic Programming vs. Greedy Local Optimization

The experiments above used greedy local placement optimization. To demonstrate the value of our dynamic programming local optimization, we compared the layout areas and trail merging for some benchmarks from the maximum density experiments above. These are tabulated in Table 1.6 . Notice that dynamic programming can achieve another 4% density improvement on average. Also, the number of trails reduced on average by 13.1% compared to the greedy local optimization. This suggests that the dynamic programming formulation is very

**TABLE 1.6** Dynamic Programming (DP) vs. Greedy local optimization

| Bench. | Layout Width Impr. DP over Greedy | No. of Trails after Greedy | No. of Trails after DP | Trail Reduction DP over Greedy | CPU Time Greedy (secs) | CPU Time DP (secs) | CPU Time DP over Greedy |
|---|---|---|---|---|---|---|---|
| frg1 | 3.7% | 118 | 96 | 18.6% | 0.18 | 0.57 | 3.2x |
| i2 | 2.4% | 116 | 108 | 6.8% | 0.25 | 0.74 | 3.0x |
| i4 | 3.9% | 191 | 169 | 11.5% | 0.28 | 0.86 | 3.1x |
| C432 | 3.4% | 226 | 197 | 12.8% | 0.31 | 0.96 | 3.1x |
| apex7 | 3.6% | 249 | 216 | 13.3% | 0.43 | 1.36 | 3.2x |
| example2 | 4.5% | 297 | 264 | 11.1% | 0.47 | 1.41 | 3.0x |
| i6 | 5.7% | 511 | 421 | 17.6% | 0.75 | 2.31 | 3.1x |

effective at detecting merges between clusters, which reduces the number of trails and also the layout area. Such an improvement is however at the expense of run-time. The dynamic programming approach took about 3 times more CPU time, on average, compared to the greedy approach.

## 4.8 Summary

Detailed placement completes our placement flow by committing each cluster to a final shape-level layout. This second phase of placement pays particular attention to detailed shape-level interactions within the clusters and between neighboring clusters, while accounting for detailed and global routability concerns. Given a global placement of clusters that meets overall wirelength, congestion and area requirements, local placement optimizes merges between neighboring clusters, while considering all feasible layout options for those clusters. Results comparing to a commercial standard cell-based layout flow demonstrate that our tool achieves 100% routed layouts that average 23% less area. In the next chapter, we describe how our flow is further enhanced to handle timing optimization during placement, to reduce overall circuit delays.

# CHAPTER 5

# Timing-Driven Placement

## 5.1 Introduction

Speed is of utmost importance to most digital circuits. Hence, reducing overall circuit delays is a goal at all stages of an ASIC flow. As technologies get smaller, the contribution from interconnect elements to the overall circuit delay increases. Optimizing for reduced circuit delays during layout synthesis is therefore very critical.

This chapter describes how we optimize for timing in our flow. In particular, the global placement of clusters most affects the lengths of individual wires, which in turn affects delays of critical paths in the circuit. Accounting for these net delays is therefore critical during global placement. We address this by enhancing our global placement phase as shown in Figure 39. The key components of our technique include:

- Transistor-level timing analysis to analyze circuit timing.
- Critical path analysis to find worst timing paths from timing analysis.
- Timing-driven placement while accounting for critical paths and nets.

**FIGURE 39.** Iterative timing-optimization during Global Placement.

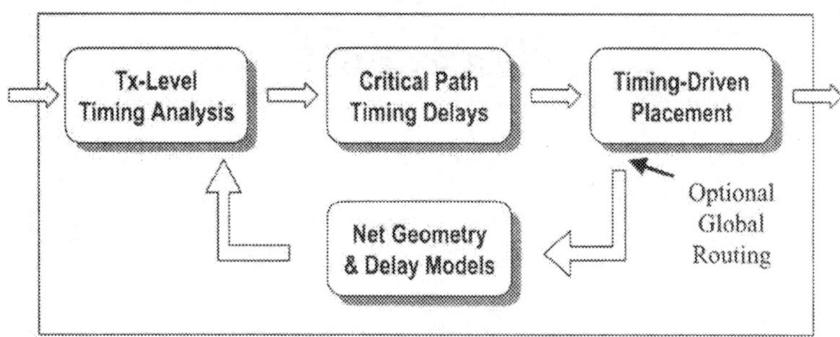


- Modeling interconnect and cluster delays.

In this chapter, we describe each of these in detail. We also present results demonstrating the effectiveness of these techniques.

## 5.2 Transistor-Level Timing Analysis

To optimize for timing during placement, we must be able to identify critical signal paths in the circuit by analyzing their timing behavior. In this section, we introduce *dc-coupled components* (DCCs) and describe how DCCs help analyze the timing behavior of flat transistor-level circuits. We review some of the standard techniques for timing analysis to motivate our choice of static timing analysis, which we describe in detail. We later describe methods for identifying timing critical paths from the results of static timing.

**FIGURE 40.** DC-Coupled Components

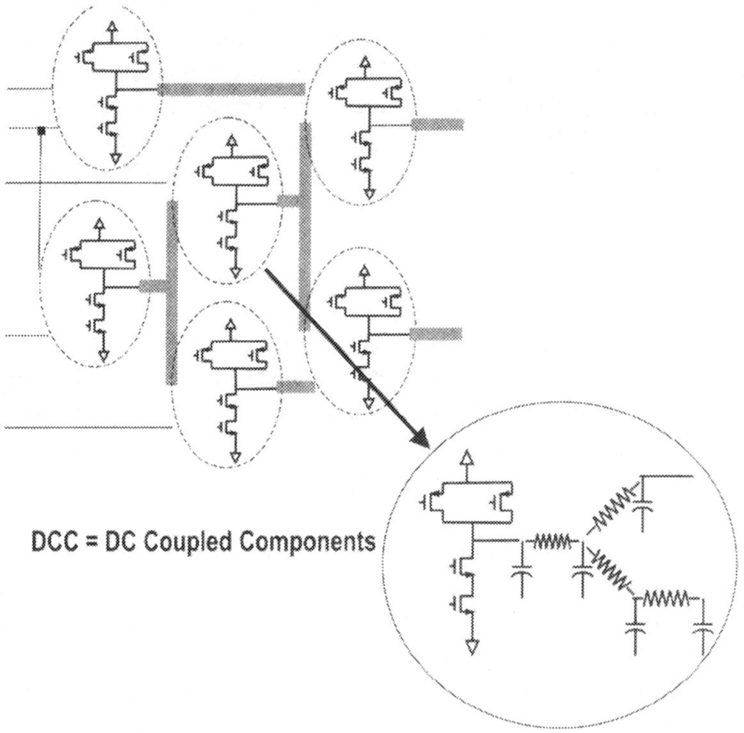

DCC = DC Coupled Components

## 5.2.1 DCCs: DC-Coupled Components

To enable timing analysis, a flat transistor-level circuit can be broken into groups of dc-coupled components or DCCs. A DCC is composed of groups of channel connected transistors and the interconnect (RC-tree model) they feed, as shown in Figure 40. This partitioning helps us separate the task of simulating signal propagation *within* these DCCs, from the task of *propagating* these signals across the circuit. In the next few sections, we describe how this is done.

Since the groups of transistors, that belong to a DCC, need to be grouped together from a timing point-of-view, it also makes sense to keep them together during placement. Since essential clusters are defined as the smallest set of transistors that need to be grouped together, clustering based on DCCs becomes essential. In our experiments therefore, there is a one-to-one correspondence between essential clusters and DCCs. However, it may be necessary to further group multiple DCCs into a single essential cluster, depending on circuit-style restrictions.

### 5.2.2 Timing Analysis

There are several standard techniques to detect timing problems in combinational logic. Broadly speaking, these techniques fall under two categories: (a) delay simulation, and (b) static timing analysis. We briefly compare the pros and cons of these two types of methods, to motivate our choice of static timing analysis.

Delay simulation, also known as dynamic timing analysis, involves simulation of the behavior of a circuit, given the input patterns at the various primary inputs. This is useful for functional and timing verification of circuits. While this method is most accurate, the analysis is limited to the given set of input patterns. A complete coverage using all sets of input patterns can be impractical for large circuits. Static timing analysis, on the other hand, tries to account for all possible input patterns at primary inputs, simultaneously. It then propagates the earliest and latest signal arrival times, across each logic path in the circuit. While doing so, it ignores the influence of other paths. In other words, at each logic gate, it assumes that the gating conditions at the gate can be set up such that signals can be propagated along any of the logic paths through the gate. It can therefore be completely independent of the input patterns or the function of the gate. This can fall victim to false paths, where timing problems are reported on paths even though those logic paths do not exist in reality. However, static timing analysis is conservative. The set of timing problems, detected by this analysis, includes any genuine timing problems in the circuit. The main advantage is that it is much faster, and thereby more practical, than delay simulation over all input patterns.

Static timing analysis can be path or block oriented. Path oriented methods work by propagating earliest and latest signals separately along each path in the circuit. Such a path enumeration returns useful information about critical signal paths and overall delay of the circuit. Block oriented static timing analysis, on the other hand, propagates signal arrival times forward through the logic. The latest and earliest possible arrival times are maintained on each gate input and output pin. The required arrival times are then propagated backwards to compute slacks at each timing node. This analysis provides useful information about slacks at individual gate outputs. As mentioned before, both these methods are much faster than delay simulation. A more detailed comparison can be found in [Fredrickson 97].

For timing-driven placement, the goal is to quickly determine the worst timing critical signal paths in the circuit, independent of the input patterns at primary inputs. Static timing analysis is therefore better suited for such an application. We now describe our approach to static timing analysis at transistor-level.

### 5.2.3 Transistor-Level Static Timing

In this section, we present our static timing approach, where we propagate signal arrival times and slews from primary inputs to primary outputs. This is similar to the block-oriented methods. However, the resulting delays in the circuit can be used for analyzing critical paths using either block or path oriented techniques. In this approach, the points of interest, for timing, are the DCC output nodes, DCC input nodes, primary outputs and primary inputs. The method works by computing signal *arrival times* and *slews* at the DCC output nodes, given all the arrival times and slews at each of the DCC input nodes. This propagation from the DCC inputs to the outputs is performed independent of the other DCCs or other logic paths. It assumes that any of the input signals can be propagated to the outputs, by appropriately setting up the other inputs. The DCCs are processed in a topological order, such that each DCC gets processed only after each of the DCCs driving it have been processed. This ensures that the signals are propagated from the primary inputs, forward, to the primary outputs. It also ensures that only the latest (or earliest) arrival times and the worst (or best) slews are propagated forward from each DCC output node.

As we can see from above, propagating these signals from DCC inputs to the DCC outputs is crucial to the working of this method. There are two key components involved in this process: (a) transistor-level DCC simulation: to propagate arrival times and slews from DCC inputs to outputs, and (b) transistor-level Boolean analysis: to compute sensitizing values at other DCC inputs, when propagating a signal from any given input. We now take a closer look at these two steps.

### Transistor-Level DCC Simulation

The goal here is to propagate signal properties from DCC inputs to outputs. The properties of interest to us are: early/late rise/fall signal arrival times and fast/slow rise/fall signal slews. The properties at the output are computed by simulating the worst-case and best-case signals from each of the DCC inputs. For each such simulation, the other inputs are set to their sensitizing values. This is illustrated in Figure 41. We perform this simulation at transistor-level using TETA [Dartu 98]. TETA applies principles of successive chord iterations for simulating nonlinear devices. By using simplified table look-up models for MOS transistors, TETA can perform simulations of these small circuits much faster than traditional SPICE-like simulators. This speed is critical to our application, since we perform multiple DCC simulations.

**FIGURE 41.** DCC Simulation

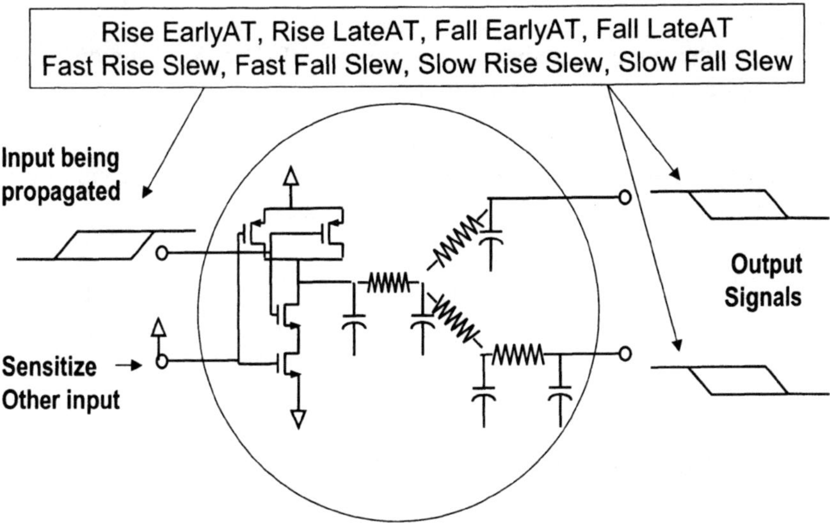


## Transistor-Level Boolean Analysis

In order to compute the sensitizing input values, when propagating from any single input, we need to compute the Boolean logic implemented by the DCC. The *unateness* of the logic at the output, with respect to the input, is also required when propagating these signals. The unateness helps us determine whether the output rises with input rise or otherwise.

For this purpose, we make use of techniques from [Bryant 87] & [Bryant 87a]. These methods use switch-level models to represent the transistor circuit. Specifically, the circuit is modeled as a network of charge storage nodes connected by resistive transistor switches. The functionality of the network is then expressed as a series of Boolean equations using a 4-valued logic to describe high, low and don't-care states. These techniques were implemented using CUDD [Somenzi 98] for all BDD-based Boolean operations. We describe the key steps involved, using an example.

Consider the circuit shown in Figure 42. The logic of the circuit is represented using the 4-valued logic as:

out.1 = a.0 + b.0, out.0 = a.1 b.1

Since we are only concerned with two-valued Boolean logic in our networks, we convert this to a 2-valued logic as:

$$out = (\overline{ab})$$

**FIGURE 42.** Boolean Analysis Example

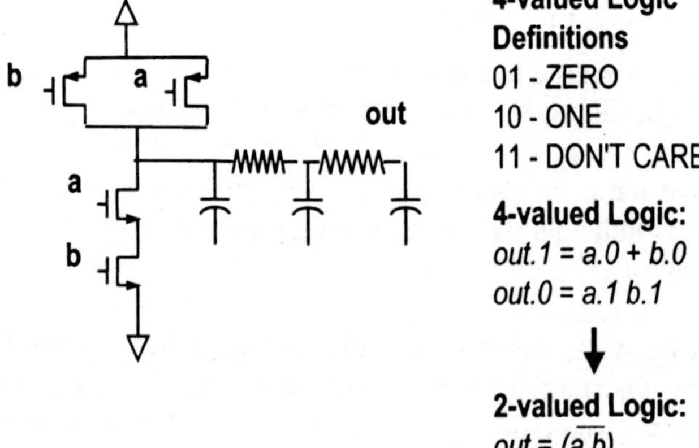

This is the more familiar NAND logic. In order to find the sensitizing input value at *b* while propagating from input *a*, we compute the Boolean derivative of the logic with respect to the propagating input and equate this to TRUE.

$$\frac{d}{da} out = Cofactor(out, a) \oplus Cofactor(out, \bar{a})$$

This yields: *b = 1* as the sensitizing value. In other words, set *b* to a high voltage. The unateness of the logic output, w.r.t. input *a*, can be computed as the cofactor of the logic, while setting other inputs to their sensitizing values.

$$Cofactor(out, \text{sensitizing\_input}) = Cofactor(out, b{=}1) = \bar{a}$$

This implies that the output falls as input *a* rises.

By combining the results of Boolean analysis with DCC simulations, we propagate the signal arrival times and slews across the DCCs, and ultimately, across the circuit, from the primary inputs to the primary outputs. In the process, we also keep track of the worst-case and best-case input-to-output delays for each of the DCCs. In the next section, we describe how we use this information to find timing critical paths in the circuit.

## 5.3 Delay Graphs and Critical Paths

Static timing analysis, described in the previous section, gives us the signal arrival times at each of the nodes in the circuit. The DCC simulation, performed during this analysis, also gives us the worst-case and best-case delays from each of the DCC inputs to the DCC outputs. In this section, we discuss path and block oriented techniques to find timing critical paths using this information.

**FIGURE 43.** Delay Graph Example

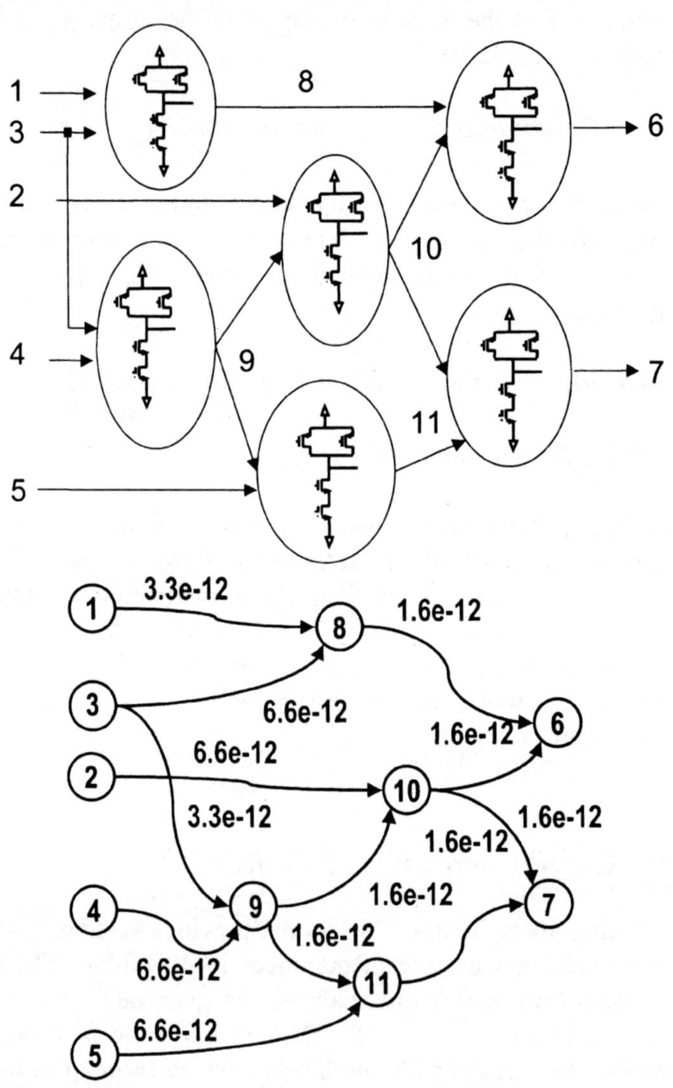

We begin by constructing a *Delay Graph* that contains a node for each DCC output, primary input and primary output. For simplicity, we only consider the worst-case delays in the circuit. The edges in the graph then represent the worst delays from each of the DCC inputs to the DCC output. Figure 43 shows an example with DCCs (above) and the corresponding Delay Graph (below). For the sake of simplicity again, interconnect is not considered here. Interconnect delays are discussed in detail in a later section. The goal now is to determine the most timing critical paths in this Delay Graph. These are the paths with the worst delays. Instead of enumerating all such paths, which can be impractical, path-oriented approaches like [Benkoski 87] use a variant of depth first analysis to detect the top critical paths. Block-oriented methods like [Hathaway], on the other hand, detect these problems by using the arrival times and computing slacks based on required arrival times.

Both the methods ([Benkoski 87] & [Hathaway]), begin by augmenting the Delay Graph to include 0-delay edges from a common source-node to all primary inputs and from all primary outputs to a common sink-node. This is illustrated in Figure 44.

[Hathaway] uses information about the Arrival Times (AT) at each of the nodes, from the forward propagation during static timing analysis. A second propagation, from the primary outputs, backward, computes the Required Arrival Times (RAT) at each of the nodes. The difference between the AT and RAT is defined as the slack. A partial-path expansion strategy, using the slack values, is then used to compute the longest-paths in the graph. [Benkoski 87] on the other hand computes an "esperance" value for each node in the graph. This is a measure of the maximum delay from the node to the sink. This metric is then used to bias partial-path expansion towards the sink. In our application, we are interested in the most critical paths. Either of these methods can therefore be used, as they end up having the same algorithmic complexity, when used for path reporting. For illustration purposes,

**FIGURE 44.** Critical Path Analysis Example

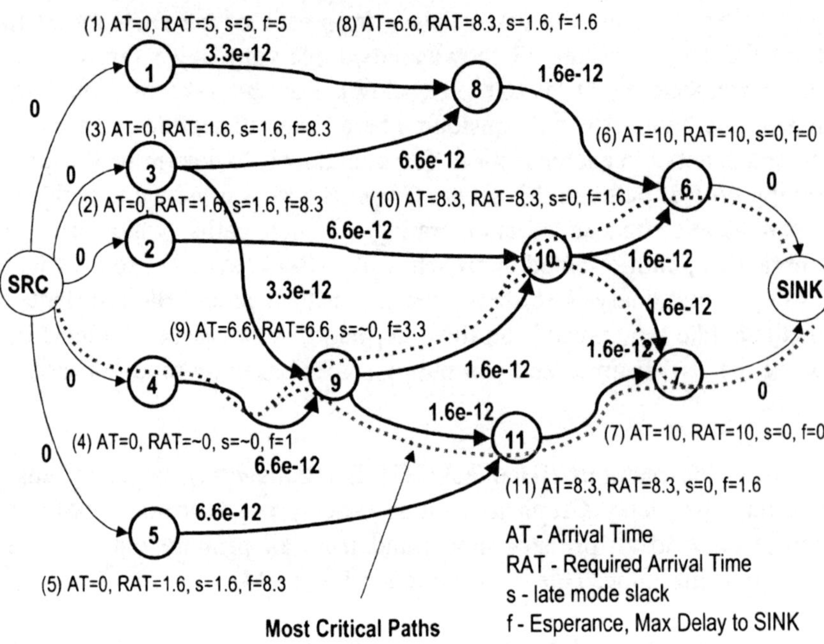


Figure 44 displays the arrival times, required arrival times, slacks and esperance values for this example, using both the techniques above. However, for our experiments, we only use the [Benkoski 87] technique. The resulting most critical paths are also shown in the figure.

## 5.4 Timing-Driven Global Placement

In the last section, we looked at analyzing the timing behavior of circuits and identifying critical signal paths in the circuit. We now look at how that information is useful during placement. The overall circuit delay is composed of two elements: the delays of the signals within the DCCs and the parasitic delays of the interconnect. The placement of the DCCs determines the lengths of interconnect which directly impacts the interconnect delay. Interconnect length also has an effect on the delays of the DCCs. The goal of timing-driven placement is to take into account these aspects of placement that influence the delays.

In this section, we look at some of the standard techniques for timing-driven placement. We then describe our approach, where we use the information from static timing analysis, to reduce the lengths of wires along critical signal paths, which in turn reduces the overall delay of the circuit. We also discuss some ways of modeling the effects of these wires on the delays of DCCs.

### 5.4.1 Timing-Driven Placement Techniques

Timing optimization during placement has typically been performed in an iterative fashion. An existing placement is analyzed for timing-problems. This information is used to improve either the current placement or a subsequent placement run. These methods generally fall into two categories: path-based methods and net-based methods. Path-based methods directly optimize for reducing the lengths of critical paths during placement, whereas the net-based methods try to reduce the lengths of individual nets that belong to these critical paths. Net-based methods tend to be faster because they do not have to perform longest path analysis during placement. However, path-based methods better capture overall timing problems and have better chances of convergence. Placement algorithms like TimberWolf [Swartz 95], that are inherently iterative in nature, have adopted the path-based approach. This algorithm also enjoys the privilege of being able to explicitly account for path-lengths in the cost-function. Direct placement techniques like GORDIAN [Kleinhans 91] and Kraftwerk [Eisenmann 98] use the net-based approach. A good review of various techniques can be found in [Riess 95].

As described previously, our global placement is based on a two-step approach. The first step uses quadratic placement & min-cut partitioning. The second step is an iterative improvement technique based on simulated annealing. We have therefore adopted a net-based approach during the first step followed by a combination of net & path-based approaches during the second step. For the net-based techniques, net-weights are computed based on the critical paths. In the next section, we describe how this is done such that it also aids timing convergence. We then describe how these weights are used within our placement algorithms.

### 5.4.2 Net-weights from Critical Paths

We have adopted the iterative net weighting scheme from [Eisenmann 98]. Based on the information about critical paths from static timing analysis, a net is critical if it belongs to a critical signal path. The criticality of net $j$ during iteration $m$ is defined as

$$c_j^{(m)} = \begin{cases} \left( c_j^{(m-1)} + 1 \right)/2 & \text{if net } j \text{ in 3 percent most critical nets} \\ c_j^{(m-1)}/2 & \text{otherwise} \end{cases}$$

Criticality is a measure of how critical the net is, depending on which critical paths the net belongs to. Using criticality, weights for net $j$ are computed as

$$w_j^{(m)} = w_j^{(m-1)} \times (1 + c_j^{(m)})$$

$$w_j^{(0)} = 1$$

Notice that both weights and criticality at iteration *m* take into account weights and criticality at iteration *m-1*. This technique aids in convergence by reducing oscillations in weights from one iteration to the next. In the next section we describe how these weights are incorporated into our placement algorithms.

### 5.4.3 Net-Weight / Timing-Driven Placement

During quadratic placement, the forces between the various essential clusters are scaled by the values of net-weights. This results in a net-weight driven quadratic solve, by using the parameter *net_wt* (see Section 3.2). Bi-partitioning also includes these net-weights during min-cut optimization. Finally, iterative improvement, via simulated annealing, includes these net-weights in the cost function (see Section 3.3). While we do not re-compute longest paths during placement optimization, we introduce a new term in the cost function, namely *PathCost*, to account for the sum of the lengths of the most critical paths.

Overall timing optimization happens in an iterative fashion. We start with a non-timing driven placement where we optimize primarily for area, wirelength and congestion. All the net-weights are initialized to a value of 1. At the end of this first iteration, we get our initial global placement, which is used to estimate interconnect lengths and delays. We then use this information to perform static timing analysis to find the critical signal paths and generate net-weights and path-information for our next iteration. In the next section, we describe how we estimate interconnect delays at the end of each global placement iteration, for use in the subsequent iteration.

### 5.4.4 Interconnect Delays

At the end of each global placement iteration, the locations of DCCs are used to generate geometric models for the interconnect. As illustrated in Figure 45, we can either use a star model for this intercon-

**FIGURE 45.** Interconnect Geometry and Models

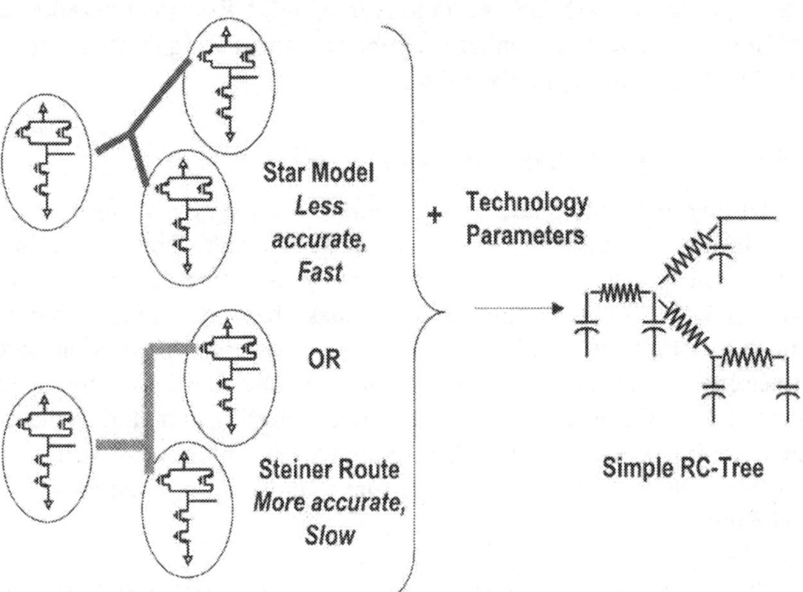



nect, or use a more accurate Steiner model. The star model is faster to compute (can be done in linear time), whereas the Steiner route takes longer. We combine this geometric model with necessary technology parameters to generate an RC-tree model. In the next section, we describe how we use this model within timing analysis, for subsequent timing-driven placement iterations.

### 5.4.5 DCC Delay Macro-Modeling

Figure 46 summarizes the timing optimization flow discussed so far. To complete the timing loop, information from the global placement result is used to generate RC-tree models for interconnect, which is then used for static timing analysis during the next iteration. One way of doing this is to include the RC-tree directly, within each DCC simu-

**FIGURE 46.** Timing Optimization Summary

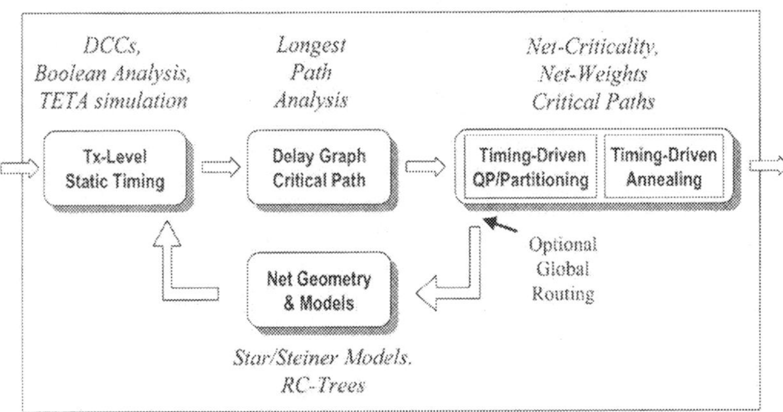


lation, in TETA. We can then perform a full simulation, by simulating each DCC, while propagating the signal arrival times and slews. This full simulation uses the correct signal slews for each DCC simulation. This also takes into account the effect of interconnect on DCC delays. However, such a full simulation can be very time-consuming, especially within multiple timing iterations. In this section, we describe a slightly different approach, that significantly reduces this time.

To reduce the number of DCC simulations and re-use as much information as possible from the fewest simulations, we resort to some simple macro-modeling techniques. These are based on approximations intended to *de-couple* DCC transistor and interconnect delays. Such a de-coupling permits us to characterize the DCC transistor delays based on the loads, thereby letting us estimate their delays without the need for further simulation.

Our basic idea of de-coupling DCC transistor delays from interconnect delays is illustrated in Figure 47. By ignoring the resistive effects

**FIGURE 47.** De-coupling DCC transistor and interconnect delays

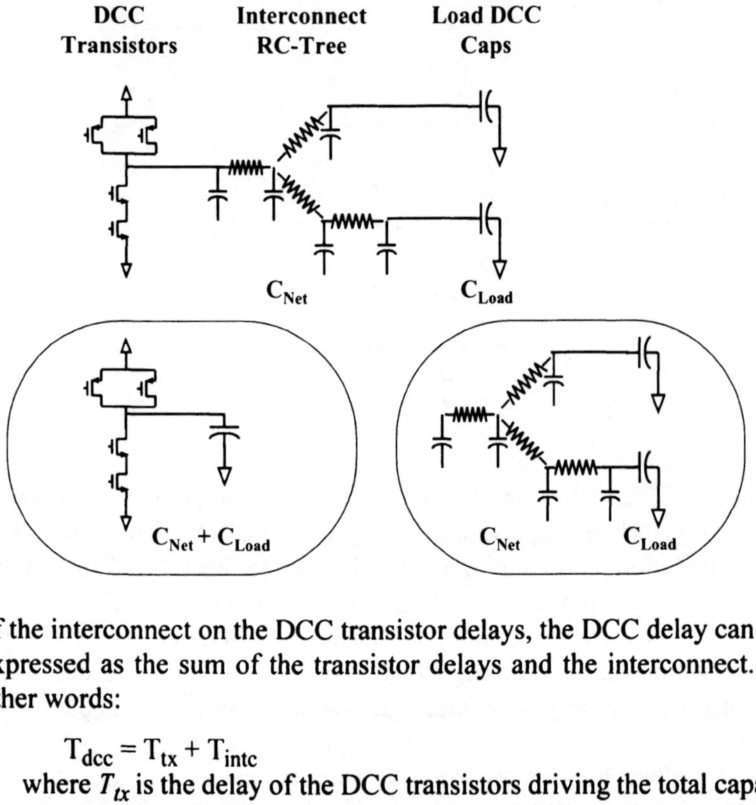

of the interconnect on the DCC transistor delays, the DCC delay can be expressed as the sum of the transistor delays and the interconnect. In other words:

$$T_{dcc} = T_{tx} + T_{intc}$$

where $T_{tx}$ is the delay of the DCC transistors driving the total capacitive load of the interconnect and the load DCCs, and $T_{intc}$ is the interconnect delay which can be estimated using elmore delay. For DCCs with identical transistor netlists, the transistor delay, $T_{tx}$ can then be characterized, based on the capacitive load, as suggested in [Weste 93]:

$$T_{tx} = T_{intrinsic} + K * C_L$$

where $T_{intrinsic}$ and $K$ are computed empirically by simulating the DCC transistors using TETA with no load modeled and then a few additional simulation "samples" with varying values of $C_L$. The values of $C_L$ are chosen based on the average loading capacitance seen by the DCC. Note that DCCs with identical transistors are clusters with the same circuit structure and the same transistor sizes. We can therefore exploit circuit structure recognition for divide-and-conquer. We have already shown that such unique structures are typically very few in number. This is where a one-to-one correspondence between DCCs and essential clusters helps. One could also consider incorporating transistor sizes into the macro-model so that we only have to simulate the essential circuit structures.

Macro-modeling described in this section makes DCC delay evaluations significantly faster. The macro-models can not only be used across multiple DCCs in the netlist, but also during multiple timing iterations. This however compromises the accuracy of such a delay evaluation by excluding interconnect and also assuming similar signal slews at each DCC. An obvious extension here is to incorporate signal slews into the macro-model. At this rather early point in the flow, the primary focus is to capture the *right* nets and paths for timing optimization. Hence, this rather simple timing model, and the decoupling of device and wiring delays in the style of a classical load-factor model, is a reasonable expedient.

In the next section, we describe experimental results demonstrating the usefulness of our techniques.


## 5.5 Timing Results

To demonstrate the effectiveness of timing optimizations on reducing the overall delays of the circuit, we compared global placement results (with global routing) with and without these optimizations. Due to the availability of some technology parameters and infrastructure, we have used STMicroelectronics 0.18um technology. The benchmarks were synthesized using lib-2 from Chapter 2.

For estimating interconnect parasitic delays, we assumed a capacitance per length of 120 pF/m and resistance per length of 867 k$\Omega$/m (using also some estimates for local/global interconnect and vias). For a typical 2-input NAND circuit simulated using TETA, the max-delay ($T_{intrinsic}$) without any load was 35 ps, for one input. Using a load capacitance of 6.2 fF, the max-delay was 55 ps, yielding a characteristic $K$ value of about 3,200.

### 5.5.1 Delay Improvement Results

In Table 1.7 , we compare timing delays for the various benchmarks with and without timing-driven optimization. Delays after each of two timing iterations are compared to the delays after placement *without* timing optimization. Note that by using our timing optimization techniques, the benchmarks showed an average delay improvement of about 10% after the first timing iteration and about 10.4% after the second timing iteration. Notice that most of the benchmarks converged to a delay value after the first timing optimization, with delays only changing by a few percent (increase/decrease) thereafter. However, some benchmarks like i9 showed significant improvement even in the second timing iteration. We observe that this behavior is related in part to the structure of the netlist, since our technique must rely on capturing the *right* critical paths. We look into this in the next section. We also notice -- unsurprisingly -- that the largest circuit, C7552, shows the largest timing improvement. As a practical matter, the smaller designs with only a few hundred transistors are really device-delay dominated. As long as we place them well, they perform well. But for the large design, there is enough "random wiring" that a poor placement with too many long paths can materially degrade the timing. Here, the timing-driven optimization makes its most visible impact.

**TABLE 1.7** Delay Improvement with Timing-driven optimization

| Bench. | No. of Trans | Delay Without Timing (ns) | Delay Timing-Iteration -1 (ns) | % Improve after Timing Iteration -1 | Delay Timing-Iteration -2 (ns) | % Improve after Timing Iteration -2 |
|---|---|---|---|---|---|---|
| frg1 | 384 | 0.774 | 0.710 | 8.3% | 0.712 | 8.0% |
| i2 | 624 | 0.450 | 0.408 | 9.3% | 0.415 | 7.8% |
| i4 | 692 | 0.652 | 0.617 | 5.3% | 0.625 | 4.1% |
| C432 | 578 | 1.553 | 1.377 | 11.3% | 1.375 | 11.5% |
| example2 | 998 | 0.862 | 0.816 | 5.3% | 0.803 | 6.8% |
| i6 | 1518 | 0.647 | 0.589 | 8.9% | 0.592 | 8.5% |
| i9 | 2000 | 1.297 | 1.172 | 9.6% | 1.105 | 14.8% |
| C7552 | 7132 | 4.190 | 3.255 | 22.3% | 3.280 | 21.7% |

Finally, we note that all the experiments execute in reasonable time. The largest benchmark with 7132 transistors took only about 15 mins to place and global route, with two timing iterations, on a desktop work-station.

### 5.5.2 Netlist Structure and Path Distribution

In Figure 48, we have drawn the delay graphs generated for two distinctly different types of benchmarks, in order to give us some insight into the structure of the netlist. Notice that the circuit C432 (left) is an example of random logic, where the netlist has reasonable depth and breadth. The i9 netlist (right), on the other hand, is broader than it is deeper, in addition to being a little more structured.

**FIGURE 48.** Netlist structure for C432 (left) and i9 (right), plotted via logic depth (inputs at left, outputs at right, in each plot)

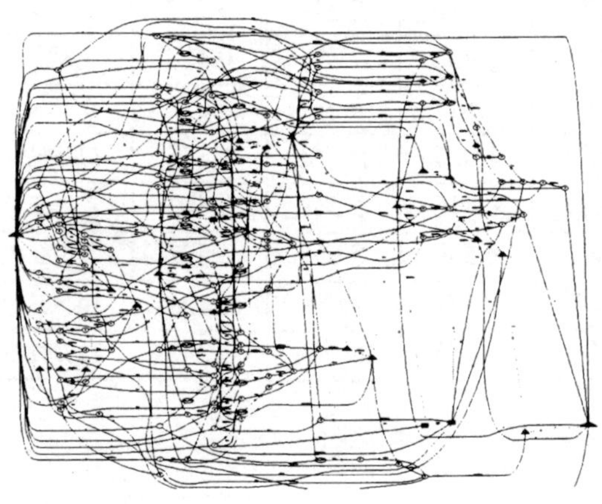

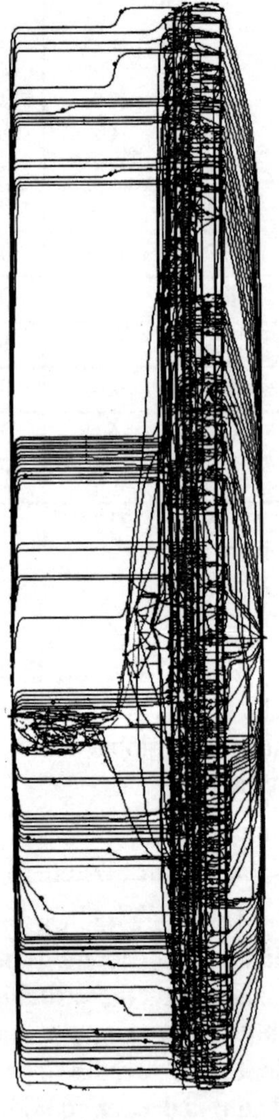

**FIGURE 49.** Path distribution for C432 before and after first timing iteration

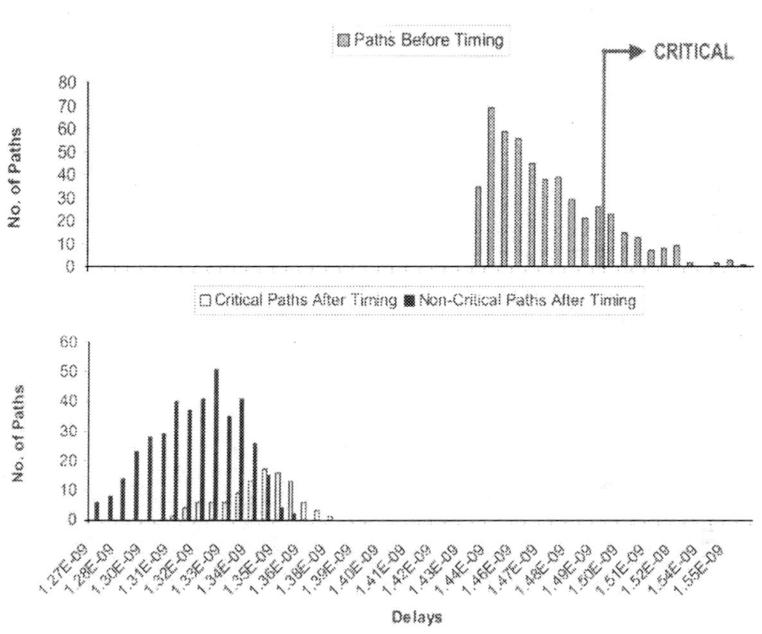



Our method relies on capturing the right critical paths from a given run and tries to reduce their lengths in the subsequent timing run. Specifically, our experiments look for the top 100 critical paths. This works well for a circuit like C432, where the top 100 paths capture a majority of the bad timing paths. We have tried to show this in Figure 49, where we analyze the top 500 timing paths in the circuit and see where the top 100 critical paths ended up, after a timing iteration, compared to the other 400 non-critical paths. Notice that for C432 (Figure 49), reducing the overall delays for the top critical paths also reduces delays for *all* the other paths.

**FIGURE 50.** Path distribution for i9 before and after first timing iteration

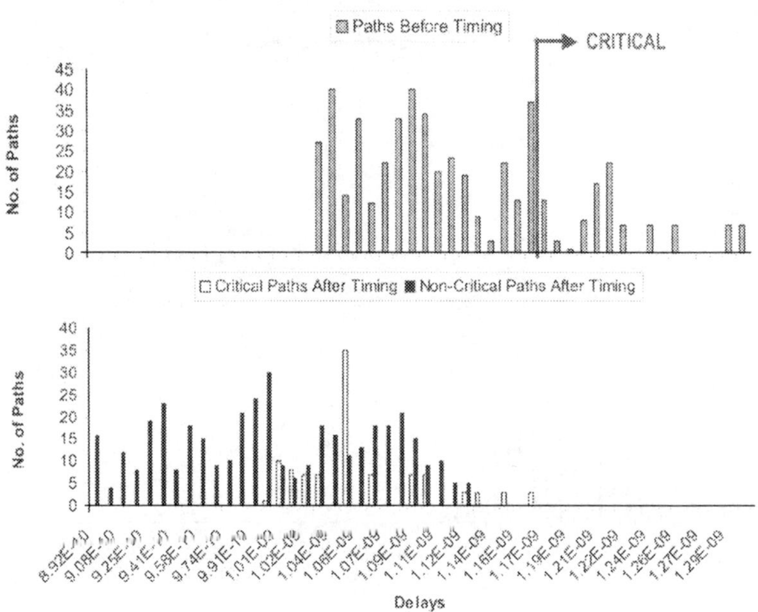


On the other hand, for the same experiment with circuit i9 (Figure 50), we notice that while the delays of the top critical paths reduced significantly, the delays of some non-critical paths did not reduce. Worse some non-critical paths from the previous run became critical for the next iteration. This is the reason we see significant improvement in the second iteration, as our net-weights in the second iteration account for the new critical paths, in addition to the older ones. This clearly shows that the *right* number of critical paths to consider and the number of timing iterations required for convergence is highly dependent on the benchmark.

## 5.6 Summary

By dividing the netlist into DCCs, we have presented an iterative timing-driven global placement strategy. We presented static timing analysis at transistor-level that makes use of transistor-level Boolean analysis and delay simulation to analyze the timing behavior of the circuit. Both path and block oriented techniques can then be used to determine the worst critical paths in the circuit. These critical paths are then optimized for, in the next placement iteration, via net weight-based and path-based placement techniques. For faster timing analysis, we use some macro-modeling techniques to approximate DCC delays and decouple DCC transistor and interconnect delays. The results show that by using our techniques, we can reduce overall circuit delays by about 10% on average.

In the next chapter, we conclude by summarizing our research contributions and identifying some areas for future research.

   *Conclusion*

We have presented a novel transistor-level layout flow, for blocks of combinational logic up to a few thousand transistors in size. Our key research contributions include:

- **New flow**: An alternative flow for designers to get directly to layout from a flat transistor-level netlist.

- **Minimal essential clustering**: Early identification of essential diffusion-merged MOS device *clusters*, but deferred binding of the clusters to a specific shape-level layout, until the very end of placement. We use pattern recognition for scalability.

- **Effective two-phase placement**: A scalable placement approach where global placement handles block-level concerns for area, wire-length, congestion and timing, while detailed shape-level concerns are addressed by a local optimization phase. The detailed placement phase accounts for detailed and global routability concerns, while generating dense layouts via inter-cluster diffusion merges.

**FIGURE 51.** Our direct transistor-level layout flow.

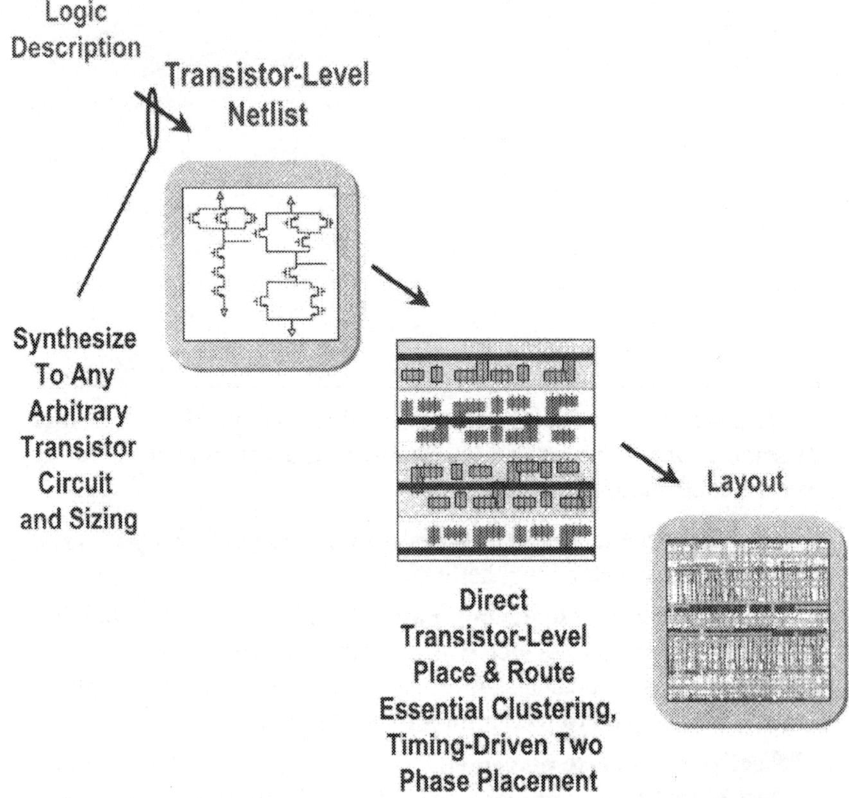


**Timing-driven placement**: We also presented timing-driven placement for transistor-level layout using transistor-level static timing analysis. The lack of such timing optimization has been a major obstacle to the adoption of previous transistor-level approaches.

A commercial router completes our flow. Experiments comparing to a commercial standard cell-level layout flow show that, when flattened to transistors, our tool consistently achieves 100% routed layouts that average 23% less area. Our approach shows that a good divide-and-conquer attack need not handicap the shape-level optimizations that have always been the distinguishing features of the best custom device-level layouts. Our emphasis on recognized circuit clusters means that natural groups of a few connected devices always have the optimal geometric arrangement. Global placement using only these clusters reduces the complexity of the problem and makes timing optimization feasible. Detailed placement focusing on intra-cluster alternatives and inter-cluster merges allows us to do crucial shape-level optimizations, but scale to large netlists. We have also demonstrated that by using our timing-driven placement techniques, the overall circuit delays can be improved by about 10-20%.

Of course, there are still several open questions remaining to be addressed, as a result of the work we have done to date, and presented here:

- **Compaction**: In order to complete the flow in an industrial environment, a compaction pass will be necessary at the end to convert our symbolic transistor placement and routing information to a usable GDSII layout. It remains to be see whether there are additional shapes-level optimizations that should be handled exclusively symbolically, with compaction as a final "legalizing" pass, or if instead some of these optimizations should just be assigned to the compactor exclusively.

- **Coupling And Noise-Aware Layout**: While our techniques work well for static CMOS, in order to make them useful for high performance circuits, placement and routing will have to consider coupling interactions and noise effects. For example, dynamic logic circuit layouts need particular attention to reduce noise effects from capacitive coupling between switching nets and sensitive nodes.

- **Role of Logic Synthesis**: In our flow, the target libraries are an artificial and intermediate construct, intended only to guide the creation of a well-structured transistor-level netlist. Given the ability to generate layout from a flat transistor-level netlist, the right target library to use during logic synthesis is an open question.

- **Sizing, Tuning and Placement**: Given recent developments in the area of flat transistor-level device sizing and tuning to meet timing and other performance concerns, a simultaneous optimization technique is worth considering.

- **ECO Placement:** To be useful in an industrial environment, incremental re-placement during ECO changes becomes a necessity. The partition of our approach into a sequence of global and local layout optimizations would seem to make the insertion of ECO optimizations quite feasible.

# *Appendix*

## A. Synthesis Target Library Subsets

For the synthesis target library comparison experiments in Chapter 2, we used some arbitrarily chosen subsets of an industrial standard cell library. In this section, we list in detail the various elements of each of those subset libraries. Note that lib-4 is the entire library, which we do not list.

Subset 1:

| Library | Elements |
|---------|----------|
| lib-0 | IVLL, ND2LL, ND3LL, ND4LL, NR2LL, NR3LL, NR4LL |

Subset 2:

| Library | Elements |
| --- | --- |
| lib-1 | **lib-0 elems** + AN2LL, AN3LL, AN4LL, OR2LL, OR3LL, OR4LL, AO10LL, AO12LL, AO13LL, AO14LL, AO15LL, AO16NLL, AO17NLL, AO20LL, AO24LL, AO2ALL, AO3ALL, AO4ALL |

Subset 3:

| Library | Elements |
| --- | --- |
| lib-2 | **lib-0 elems** + AN2LL, AN2LLP, AN2LLX3, AN2LLX4, AN3LL, AN3LLP, AN3LLX3, AN3LLX4, AN3LLX8, AN4LL, AN4LLP, AN4LLX3, AN4LLX4, IVLLP, IVLLX05, IVLLX16, IVLLX3, IVLLX32, IVLLX4, IVLLX5, IVLLX8, ND2LLP, ND2LLX05, ND2LLX3, ND2LLX4, ND3ALL, ND3ALLP, ND3ALLX3, ND3ALLX4, ND3LLP, ND3LLX05, ND3LLX3, ND3LLX4, ND4LLP, ND4LLX05, ND4LLX3, ND4LLX4, NR2LLP, NR2LLX05, NR2LLX3, NR2LLX4, NR3ALL, NR3ALLP, NR3ALLX3, NR3ALLX4, NR3LLP, NR3LLX05, NR3LLX3, NR3LLX4, NR4LLP, NR4LLX05, NR4LLX3, NR4LLX4, NR4LLX8, OR2LL, OR2LLP, OR2LLX3, OR2LLX4, OR2LLX8, OR3LL, OR3LLP, OR3LLX3, OR3LLX4, OR4LL, OR4LLP, OR4LLX3, OR4LLX4 |

Subset 4:

| Library | Elements |
| --- | --- |
| lib-3 | **lib-2 elems +** AO10LL, AO10LLX05, AO10NLL, AO10NLLP, AO11LL, AO11LLP, AO11LLX05, AO11NLL, AO11NLLP, AO12LL, AO12NLL, AO12NLLP, AO13LL, AO13LLX05, AO13NLL, AO13NLLP, AO14LL, AO14NLL, AO14NLLP, AO15LL, AO15LLX05, AO15NLL, AO15NLLP, AO16LL, AO16NLL, AO16NLLP, AO17LL, AO17LLX05, AO17NLL, AO17NLLP, AO18LL, AO18LLX05, AO18NLL, AO18NLLP, AO1ALL, AO1ALLP, AO1ANLL, AO1ANLLP, AO1ANLLX4, AO1CLL, AO1CLLP, AO1CLLX4, AO1CNLL, AO1CNLLP, AO1CNLLX4, AO1LL, AO1LLP, AO1LLX05, AO1NLL, AO1NLLP, AO1NLLX4, AO20LL, AO20LLX05, AO20NLL, AO20NLLP, AO21LL, AO21LLP, AO21LLX05, AO21NLL, AO21NLLP, AO21NLLX4, AO22LL, AO22LLX05, AO22NLL, AO22NLLP, AO23LL, AO23LLX05, AO23NLL, AO23NLLP, AO24LL, AO24LLX05, AO24NLL, AO24NLLP, AO25LL, AO25NLL, AO25NLLP, AO26LLX05, AO26NLL, AO26NLLP, AO27LLX05, AO27NLL, AO27NLLP, AO2ALL, AO2ALLP, AO2ALLX4, AO2ANLL, AO2ANLLP, AO2ANLLX4, AO2LL, AO2LLP, AO2LLX05, AO2NLL, AO2NLLP, AO2NLLX4, AO39LL, AO3ALL, AO3ALLP, AO3ANLL, AO3ANLLP, AO3ANLLX4, AO3CLL, AO3CLLP, AO3CNLL, AO3CNLLP, AO3CNLLX4, AO3LL, AO3LLP, AO3LLX05, AO3NLL, AO3NLLP, AO3NLLX4, AO40LL, AO40LLP, AO41LL, AO41LLP, AO4ALL, AO4ALLP, AO4ANLL, AO4ANLLP, AO4ANLLX4, AO4LL, AO4LLP, AO4LLX05, AO4NLL, AO4NLLP, AO4NLLX4 |

## B. Netlist Structure

For each of the circuits we used in our experiments in Chapter 5, we show here the netlist plotted via logic depth, inputs on the left and outputs on the right. Notice that netlists come in a variety of shapes, with varying logic breadth and depth. While some of these are examples of random logic, other netlists are more structured.

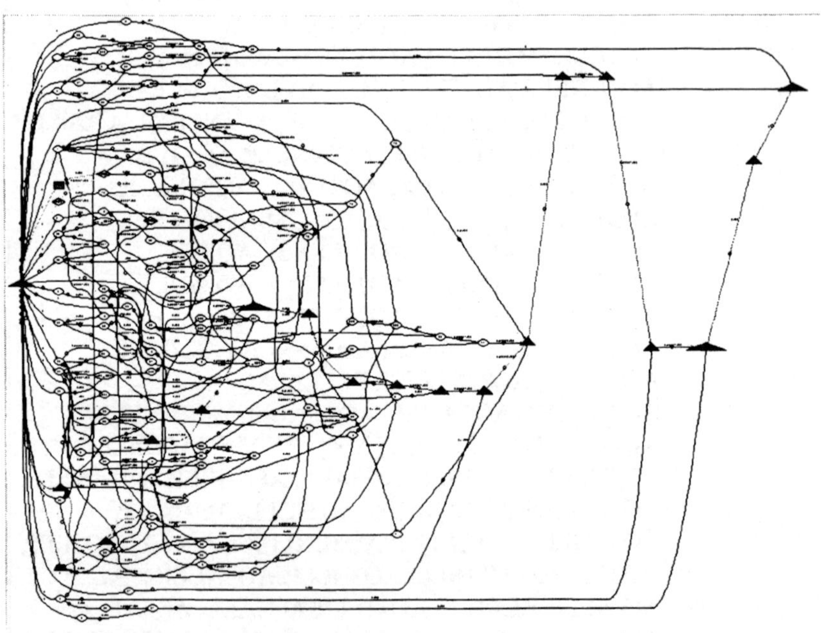

Circuit frg1, 384 transistors

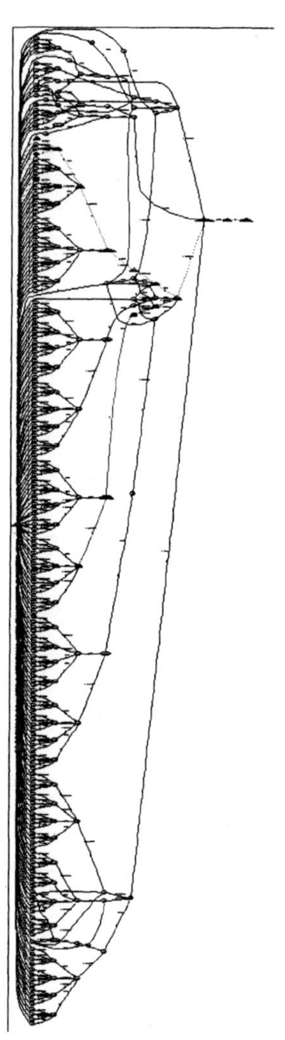

Circuit i2, 624 transistors

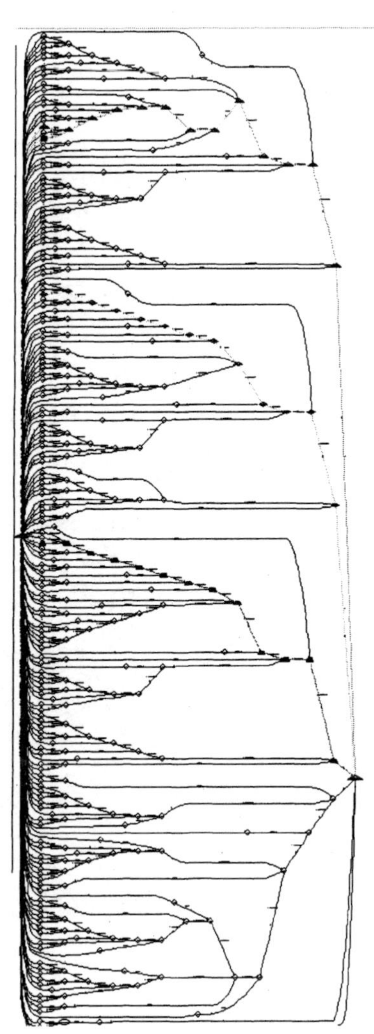

Circuit i4, 692 transistors

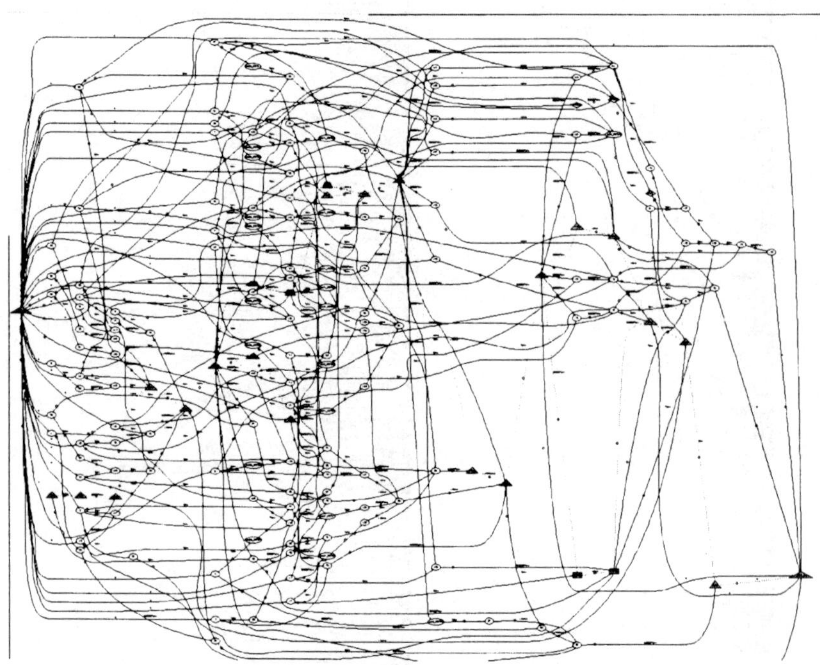

Circuit C432, 578 transistors

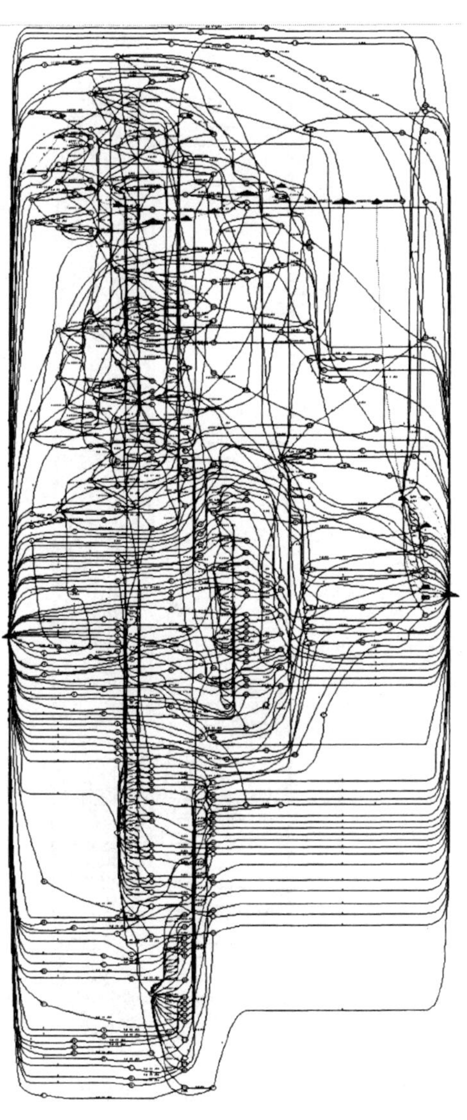

Circuit example2
998 transistors

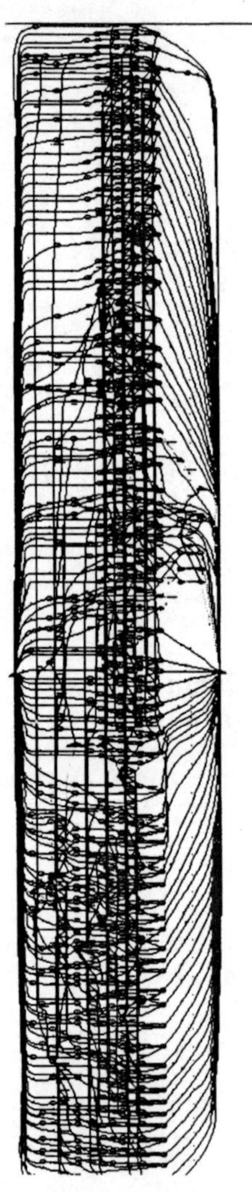

Circuit i6, 1518 transistors

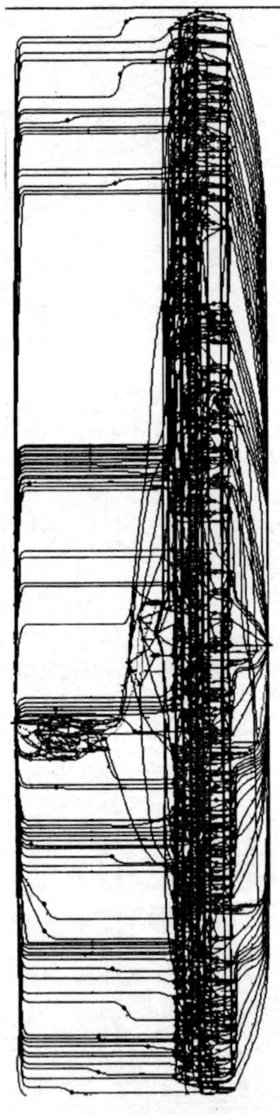

Circuit i9, 2000 transistors

# *Bibliography*

[Basaran 97]   B. Basaran, "Optimal Diffusion Sharing in Digital and Analog CMOS Layout," Ph.D. Dissertation, Carnegie Mellon University, CMU Report No. CMUCAD-97-21, May 1997.

[Benkoski 87]   J. Benkoski, E. Vanden Meersch, L. Claesen, and H. De Man, "Efficient Algorithms for Solving the False Path Problem in Timing Analyzers," *IEEE-ICCAD Digest of Technical Papers*, 1987.

[Bryant 87]   R. E. Bryant, "Algorithmic Aspects of Symbolic Switch Network Analysis," *IEEE Transactions on Computer-Aided Design of Integrated Circuits and Systems*, CAD-6, No. 4, July 1987, pp. 618-633.

[Bryant 87a]   R. E. Bryant, "Boolean Analysis of MOS," *IEEE Transactions on Computer-Aided Design of Integrated Circuits and Systems*, CAD-6, No. 4, July 1987, pp. 634-639.

| | |
|---|---|
| [Burns 98] | J. Burns and J. Feldman, "C5M-A Control Logic Layout Synthesis System for High-Performance Microprocessors," in proc. 1997 ISPD, pp. 110-115. |
| [Cadence] | http://www.cadence.com |
| [CBL] | http://www.cbl.ncsu.edu/benchmarks |
| [Chow 92] | S. Chow, H. Chang, J. Lam, and Y. Liao, "The Layout Synthesizer: An Automatic Block Generation System," in proc. *1992 CICC.*, pp. 11.1.1-11.1.4. |
| [Cohn 94] | J. M. Cohn, D. J. Garrod, R. A. Rutenbar and L. R. Carley, "Analog Device-Level Layout Automation," Kluwer Academic Publishers, Boston MA., 1994. |
| [Dartu 98] | F. Dartu, L. T. Pileggi, "TETA: Transistor-Level Engine for Timing Analysis," in Proc. *35th DAC*, June 1998, pp. 595-598. |
| [Detjens 98] | E. Detjens, G. Gannot, R. Rudell, A. Sangiovanni-Vincentelli and A. Wang, "Technology mapping in MIS," Proc. Int. Conf. CAD (ICCAD- 87), pp. 116-119, Nov. 1987. |
| [Eisenmann 98] | H. Eisenmann and F. Johannes, "Generic Global Placement and Floorplanning," Proc. Design Auto. Conf., June 1998. |
| [Fredrickson 97] | M. S. Fredrickson, "Static Timing Analysis - 101," IBM Internal Report, 1997. |

[Guan 95]      B. Guan and C. Sechen, "An area minimizing layout generator for random logic blocks," *Proc. 1995 IEEE Custom IC Conf.*, May 1995, pp. 23.1/1-4.

[Guan 96]      Bingzhong Guan, Carl Sechen, "Large Standard Cell Libraries and Their Impact on Layout Area and Circuit Performance," ICCD 1996: 378-383.

[Gupta 99]      A. Gupta and J. Hayes, "Near-Optimum Hierarchical Layout Synthesis of Two-Dimensional CMOS Cells," *Proc. 12th International Conference on VLSI Design,* Jan 1999, pp. 453-459.

[Guruswamy 97]      M. Guruswamy, R. Maziasz, D. Dulitz, S. Raman, V. Chiluvuri, A. Fernandez and L. Jones, "CELLERITY: A Fully Automatic Layout Synthesis System for Standard Cell Libraries," in proc. 1997 DAC, pp. 327-332.

[Hathaway]      David J. Hathaway, IBM Corp., Essex Junction, VT. *Private communications.*

[Her 91]      T. W. Her and D. F. Wong, "Optimal Module Implementation and its Application to Transistor Placement," IEEE International Conference on Computer-Aided Design, 1991, 98-101.

[Hsieh 91]      Y. Hsieh, C. Hwang, Y. Lin, Y. Hsu, "LiB: A CMOS Cell Compiler," *IEEE Trans. on CAD,* 10(8), August 1991, pp. 994-1005.

[Hustin 87]      S. Hustin and A. Sangiovanni-Vincentelli. TIM, "A new standard cell placement program based on the simulated annealing algorithm," *IEEE Physical Design Workshop on Placement and Floorplanning.* April, 1987.

[Karypis 97]      G. Karypis, R. Aggarwal, V. Kumar and S. Shekhar, "hMetis, A Hypergraph Partitioning Package, Version 1.0," Manuscript, December 1997.

[Kirkpatrick 83]  S. Kirkpatrick, C. D. Gelatt, and M. P. Vecchi, "Optimization by Simulated Annealing," *Science,* Vol. 220, No. 4598, November 83, pp. 671-680.

[Kleinhans 91]    J. Kleinhans, G. Siegl, F. Johannes, K. Antreich, "GORDIAN: VLSI Placement by Quadratic Programming and Slicing Optimization," *IEEE Trans. CAD,* vol 10 no 3, March 1991.

[Krasnicki 97]    M. J. Krasnicki, "Generalized Analog Circuit Synthesis," Master's Thesis, Carnegie Mellon University, CMU Report CMUCAD-97-59, December 1997.

[Lam  88]         J. Lam and J. M. Delosme, "Performance of a New Annealing Schedule," *Proceedings of IEEE/ACM Design Automation Conference,* pages 306-311. 1988.

[Lefebvre 89]     M. Lefebvre and C. Chan, "Optimal Ordering of Gate Signals in CMOS Complex Gates," in proc. 1989 CICC, pp. 17.5.1-17.5.4.

[Lefebvre 92]     M. Lefebvre and D. Skoll, "PicassoII: A CMOS Leaf Cell Synthesis System," in proc. *1992 MCNC Intl. Workshop on Layout Synth.,* vol. 2, pp. 207-219.

[Madden]          Patrick H. Madden, "The Generic C Library," http://vlsicad.cs.binghamton.edu/~pmadden.

[Malavasi 95]    E.Malavasi, D. Pandini, "Optimum CMOS Stack Generation with Analog Constraints," *IEEE Transactions on Computer-Aided Design*, Vol. 14, No. 1, January 1995, pp. 107-122.

[Maziasz 92]    R.L. Maziasz, J.P. Hayes, *"Layout Minimization of CMOS Cells,"* Kluwer Academic Publishers, Boston/Dordrecht/London, 1992.

[McDonald 01]    C. B. McDonald and R. E. Bryant, "Symbolic Functional and Timing Verification of Transistor-Level Circuits," IEEE Transactions on Computer-Aided Design, March 2001.

[Messmer 95]    B.T. Messmer and H. Bunke, "Subgraph isomorphism in polynomial time," Technischer Bericht IAM 95-003, Institut fur Informatik, Universitat Bern, Schweiz, 1995.

[Ohlrich 93]    M. Ohlrich, C. Ebeling, E. Ginting, L. Sather, "SubGemini: Identifying SubCircuits using a Fast Subgraph Isomorphism Algorithm," Proc. *1993 DAC*, pp. 31-37.

[Riess 95]    B. M. Riess and G. G. Ettelt, "Speed: Fast and Efficient Timing Driven Placement," *Proc. Intl. Symp. on Circuits And Systems*, 1995, pp. 377-380.

[Riepe 99]    M. A. Riepe, "Transistor Level Micro Placement and Routing for Two-Dimensional Digital VLSI Cell Synthesis," Ph.D. dissertation, University of Michigan, 1999.

[Rudell 89]    R. Rudell, "Logic synthesis for VLSI design," Technical Report UCB/ERL M89/49, University of California, Berkeley, April 1989.

| | |
|---|---|
| [Rutenbar 89] | Rob A. Rutenbar, "Simulated Annealing Algorithms: An Overview," *IEEE Circuits And Devices Magazine,* Vol. 5, No. 1, January 89, pp. 19-26. |
| [Sadakane 95] | T. Sadakane, H. Nakao, and M. Terai, "A new hierarchical algorithm for transistor placement in CMOS macro cell design," Proceedings of CICC-95, pp. 461-464. |
| [Sechen 85] | C. Sechen and A. Sangiovanni-Vincentelli, "The TimberWolf Placement and Routing Package," *IEEE Journal of Solid-State Circuits,* Vol. SC-20, No. 2, April 1985, pp. 510-522. |
| [Sechen 86] | C. Sechen and A. Sangiovanni-Vincentelli, "TimberWolf 3.2: A New Standard Cell Placement and Global Routing Package," in proc. *23rd DAC,* 1986, pp. 432-439. |
| [Sentovich 92] | E. M Sentovich, K. J. Singh, L. Lavagno, C. Moon, R. Murgai, A. Saldanha, H. Savoj, P. R. Stephan, R. K. Brayton, A. Sangiovanni-Vincentelli, "SIS: A System for Sequential Circuit Synthesis," Dept. of EECS, University of California, Berkeley, 1992 |
| [Serdar 99] | T. Serdar, C. Sechen, "AKORD: Transistor-Level and Mixed Transistor/Gate Level Placement Tool for Digital Data Paths," in Proc. 1999 *ICCAD.* |
| [Skalicky 96] | Tomas Skalicky, "LASPACK," http://www.tu-dresden.de/mwism/skalicky/laspack/laspack.html. |

[Somenzi 98]   F. Somenzi, *"CUDD: CU Decision Diagram Package - Release 2.2.0, Online User Manual,"* May 1998.

[STM]   STMicroelectronics, http://www.stmicroelectornics.com/

[Swartz 90]   W. Swartz and C. Sechen, "New Algorithms for the Placement and Routing of Macrocells," *Proceedings of IEEE/ACM International Conference on Computer-Aided Design*, pages 336-339. November, 1990.

[Swartz 95]   W. Swartz, C. Sechen, "Timing driven placement for large standard cell circuits," in *ACM/IEEE DAC*, 1995.

[Synopsis]   http://www.synopsys.com/

[Tani 91]   K. Tani, K. Izumi, M. Kashimura, T. Matsuda and T. Fujii, "Two-Dimensional Layout Synthesis for Large-Scale CMOS Circuits," in proc. *1991 ICCAD*, pp. 490-493.

[Tsay 88]   R. S. Tsay, E. Kuh, C. P. Hsu, "PROUD: A Sea-Of-gates Placement Algorithm," *IEEE Design & Test of Computers*, Dec 1988.

[Weste 93]   N.H.E. Weste and K. Eshraghian, *"Principles of CMOS VLSI Design,"* Addison-Wesley Publishing Company, 1993.

# *Index*

Lightning Source UK Ltd.
Milton Keynes UK
UKOW03n2235220713

214201UK00008B/74/A